Cambridge County Geographies

SOMERSET

by

FRANCIS A. KNIGHT
assisted by
LOUIE M. (KNIGHT) DUTTON

With Maps, Diagrams and Illustrations

Cambridge:
at the University Press
1909

CAMBRIDGE UNIVERSITY PRESS
Cambridge, New York, Melbourne, Madrid, Cape Town,
Singapore, São Paulo, Delhi, Mexico City

Cambridge University Press
The Edinburgh Building, Cambridge CB2 8RU, UK

Published in the United States of America by
Cambridge University Press, New York

www.cambridge.org
Information on this title: www.cambridge.org/9781107626690

First published 1909
First paperback edition 2012

A catalogue record for this publication is available from the British Library

ISBN 978-1-107-62669-0 Paperback

PREFACE

IN compiling this volume much use has been made of the *Proceedings* of the Somersetshire Archaeological and Natural History Society, and of the first volume of the *Victoria County History of Somerset*. The writer's thanks are also due to Dr Mill and Dr Shaw for valuable information on the subject of rainfall and climate, and to many others who have kindly assisted in the work.

F. A. K.

1 *May* 1909.

CONTENTS

ILLUSTRATIONS

MAPS

I. County and Shire. The name Somerset.

The word "shire," which is derived from an Anglo-Saxon root meaning to shear or cut—hence that which is shorn or cut off, a division—was formerly of wider application than it is in our time, and was originally applied to a division of a district or even of a town. There were once, for example, six small "shires" in the county of Cornwall, and there are seven "shires" in the city of York.

Tradition wrongly assigns to King Alfred the division of all England into shires. But the process was really a gradual one, and was the work of many hands; and it was not complete even at the time of the Norman Conquest. Indeed, even now alterations are being made. The county of London has assimilated portions of others besides Middlesex, especially Essex; and, as we shall see later, parts of Somerset have recently been incorporated with surrounding counties.

The object of thus dividing up the country was partly military and partly financial. Every shire was bound to provide a certain number of armed men to fight the king's battles, and also to pay a certain sum of money towards the king's income; and in each case a "shire-reeve," or, as we call the officer now, a sheriff, was appointed by the sovereign to see that the district did its duty in both respects. The word "county" came into use after the Norman Conquest, when the government of each shire was handed over to a count, a title which originally meant a "companion" of the king.

It is usually considered that the names of those counties which end with the syllable "shire" once formed parts of larger districts or of ancient kingdoms. Those which do not end with it are believed to represent entire kingdoms or tribal divisions. Thus Yorkshire is a "shire" because it once formed part of the kingdom of Northumbria; and Kent is not a "shire" because it is practically identical with the old kingdom of the Cantii.

In the case of our own county both forms are in use. Some people prefer "Somerset," and some prefer "Somersetshire," and there is something to be said in favour of both forms. Since the county was inhabited, as authorities believe, by a single tribe, the Seo-mere-saetan, "the dwellers by the Sea-lakes," that great inland water which formerly occupied so much of its low-lying districts, and since it thus represents a distinct tribal division, it should be called Somerset. On the other hand, because it was, at a later period, separated from the kingdom of Wessex, there is ground for calling it Somersetshire. Again, "Somersete" is the spelling given in Domesday Book, which was finished in 1086. But in the Exeter Domesday, which was compiled little, if any, later than the wider and more general survey, the name of the county

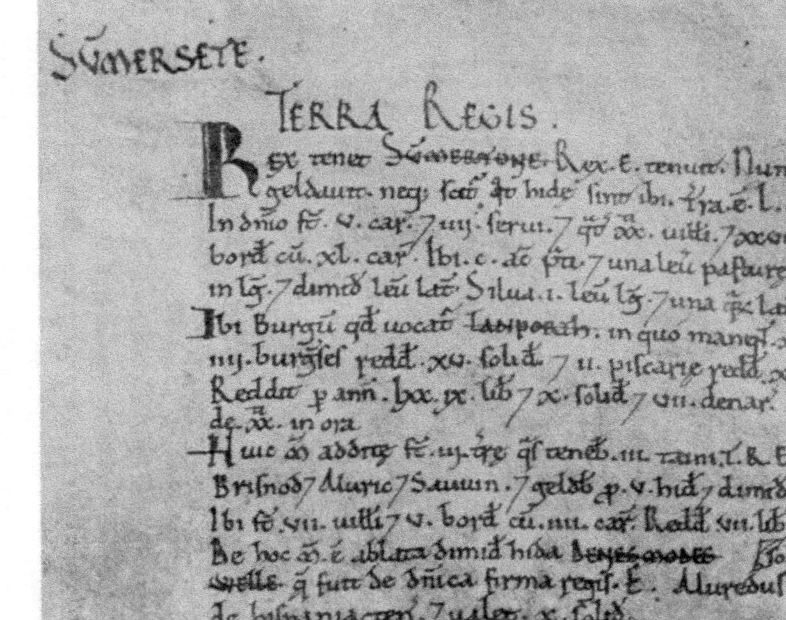
SUMERSETE.

TERRA REGIS.

Rex tenet Sumertone. Rex. E. tenuit. Nunq.
geldauit. neq. scit qt hide sint ibi. Tra. e. L. car.
In dnio st. v. car. 7 iiij. serui. 7 qt xx. uitti. 7 xxiiij.
bord cu. xl. car. Ibi. c. ac pti. 7 una leu pasturę
in lg. 7 dimid leu lat. Silua. i. leu lg. 7 una qc lat.
Ibi burgu qd uocat Lanport. in quo manet xxx
iiij. burgses redd. xv. solid. 7 ii. piscariae redd. x. sol.
Reddit p ann. lxx. ix. lib 7 x. solid 7 vii. denar.
de. xx. in ora.
Huic m addit st. iiij. trae qs teneb. iii. taini. T. R. E.
Brisnod 7 Aluric 7 Sawin. 7 geldb p. v. hid 7 dimid.
Ibi st. vii. uitti 7 v. bord cu. iiii. car. Redd vii. lib 7 xv.
De hoc m. e ablata dimid hida ~~Bennesmodes~~ solid.
~~welle~~ q fuit de dnica firma regis. E. Aluredus
de hispania ten. 7 ualet. x. solid.

Facsimile of Domesday Book (slightly reduced)

is written "Summerseta Syra," that is to say, "Somersetshire." It is "Sumersetescire" in the Anglo-Saxon Chronicle, in an entry made in 1122. It was Somersetshire to Leland and Camden in the days of Henry VIII and Elizabeth, and to writers of the seventeenth and eighteenth centuries. It is Somersetshire in the histories of Macaulay, Froude, and John Richard Green. Even Freeman, who was the great champion of the shorter form, himself frequently employed the longer one. It has been well said that the name of a thing is that by which it is commonly called. And although some of us may prefer to call our county Somerset, on the ground that that name is probably the older, and possibly the more correct, others may fairly claim that the word Somersetshire has been in use for more than eight hundred years.

In either case the name of the county is probably derived from that of the Seo-mere-saetan already alluded to, the tribe who were living here at the time of the Saxon conquest. What little we know of these primitive people will be described in later chapters. They left no written records, and we cannot tell if they had any definite organisation, or were governed by a king. Nor do we know what was their western boundary before the Saxon invasion. The Saxons are believed to have established their frontier first at the river Axe, and then at the river Parrett. It is thought that the border-line was moved further westward by King Alfred, and that it was in the reign of that monarch that the county assumed its present form.

2. General Characteristics.

Somerset is a county in the far west of England, forming part of the southern shore of the estuary of the Severn, which in consequence of the great importance of Bristol as a seaport is better known as the Bristol Channel. It is a county with a character peculiarly its own. Perhaps there is not, among all the forty shires of England, another in which there is so complete a sense of quiet and repose, of rest and peace and pleasant rural charm, as that which characterises the hills and valleys, the green orchard aisles, the broad and fertile meadows of this beautiful land. It is above everything an agricultural county, and its methods of farming have, in the course of many ages, reached a high degree of excellence. But farming, even at its best, is a pursuit which serves but to deepen the sense of rest and peace. Nowhere is Somerset the seat of any important manufacture. It does not possess even one great industrial town, with roaring mills and crowded streets and the noise and stir of hurrying traffic.

Maritime although it is, there is not, in all its sixty miles of sea-board, a single harbour worthy of the name. The ocean highway of the Bristol Channel lies far out from its shores, and the only substantial advantage that it derives from its position is in its watering-places, some of which—although the largest of them would elsewhere be called a little town—have of late years become popular as health or holiday resorts.

The coast of Somerset, much of which lies very low and some of it even below the level of high-water mark, has been formed to a great extent out of mud and sand brought down by the Severn, whose estuary is fringed in many places with broad sands, and still broader mud-flats, where shallow waters make navigation difficult and dangerous.

But the very cause which renders the maritime position of the county of little commercial importance to it, adds greatly to its prosperity in another and very different way. The alluvial lands not only lie along the shore, but extend far into the heart of Somerset. Ages of cultivation have converted what were once almost impassable morasses into some of the most fertile soil in England, and this it is which has gained for Somerset its reputation as one of the best grazing and dairy districts in the island. There is no land in Britain, for example, to surpass in richness the meadows of Taunton Dean, or the even more famous Pawlett Hams on the shore of the river Parrett.

One might well imagine, looking down from the Mendips or the Quantocks or the eastern slopes of Exmoor, that this was a district whose peace had never been broken by the clash of arms. It is, no doubt, a county of long tenures. Macaulay declared that he had found in it farmers in possession of lands that their ancestors had held in the time of the Plantagenets. The castle of Dunster has changed hands only once since its foundation by one of William the Conqueror's knights, while the manor of East Quantockshead still belongs to the family who owned it when its details were set down in Domesday Book.

And yet the record of Somerset has been a stirring one. It may be doubted if there are many shires in England of greater interest to the historian. Within its borders are some of the most famous spots in Britain; places of national importance, whose names are as familiar as those of Hastings or of Runnymede. Such, for example, is the Isle of Athelney, where Alfred paused for breathing-space before his great victory at Ethandune. Such is Sedgemoor, last of English battlefields. Then, too, of all heroic episodes in the Civil War, what is more stirring than Blake's defence of Taunton? In traces of times before the dawn of history the county is extraordinarily rich. Few English towns have yielded so many Roman remains as the city of Bath. In few other English counties are the hill-tops crowned with so many ancient strongholds. To the naturalist, the student of architecture, and the lover of beautiful scenery, the attractions of Somerset are of a very high order. It is indeed a county to be proud of, a pleasant land to live in, a region of unfailing interest and charm.

3. Size. Shape. Boundaries.

Somerset, which ranks seventh in size among the counties of England, and occupies about one-thirty-second of the area of the whole country, is irregular in shape, but may be roughly compared to three quadrants of a horizontal oval.

Its greatest length, measured from east to west along a horizontal line drawn from the cross-roads in Longleat Park on the borders of Wiltshire, to Saddler's Stone, on Exmoor, close to the edge of Devonshire, is 67 miles; and its greatest breadth, measuring due south from the entrance of the Bristol Avon, is 43 miles. It may be

Gorge of the Avon: Clifton Suspension Bridge

added that it is only in Yorkshire and in Sussex that a longer horizontal line can be drawn. Its area, which varies according to whether we are speaking of the registration county, the administrative county, or the ancient geographical county, is rather more than a million acres, or not far from 1700 square miles. Compared with the counties that adjoin it, it is a good deal smaller than Devon, but is larger than Dorset, Wiltshire, or Gloucestershire. It is ten times the size of Rutland, but is only one-third as large as Yorkshire.

Somerset has on the north two natural boundaries, the Bristol Channel and the Bristol Avon, which stream divides it from Gloucestershire. On the other sides its frontiers are more arbitrary and artificial, and depend to a less degree upon physical features. It is bounded on the east by Wiltshire and Dorset, the border being marked for a short distance in the north-east by the river Frome, and further south by the Wiltshire Downs; and on the south by Dorset and Devon, the Black Down Hills forming part of the frontier between it and the latter county. On the west it is bounded by Devon, which there shares with Somerset the great upland of Exmoor.

Some years ago, no fewer than twenty parishes, or parts of parishes which were separated from Somerset and were entirely surrounded by one or other of the adjoining shires, were still considered to belong to this county. But these outlying districts have all been handed over by Act of Parliament to the counties in which they are actually situated. Such, for example, are Poyntington, Maiden Bradley, Culmstock, and Mangotsfield, which are now included in Dorset, Wiltshire, Devon, and Gloucestershire respectively.

4. Surface and General Features.

Somerset is a beautiful county; and its varied scenery, its hills and valleys, and its wide stretches of level and low-lying plain, each with its own particular charm and its own historic and antiquarian interests and associations, are among its principal and most valued possessions.

The impression left on the mind of the traveller who crosses the county by the main line of the Great Western Railway, the railway that joins Bristol with Exeter, and London with the far extremity of Cornwall, will probably be of a flat and featureless region, with few towns, and with scattered hamlets and solitary farmsteads dotted at distant intervals over the green levels of a far-reaching and monotonous plain. Such, indeed, is the character of the turf moors that occupy so much of the north and centre of Somerset, and form one of the special features of the county. Yet Somerset has more of hill than of plain, and it contains some of the finest scenery in the West of England.

The general surface of the county admits of three broad divisions. The north, which is mainly occupied by the Mendips and their more or less connected outlying ranges, is for the most part hilly and undulating. The central region is so flat, and its lines of hill so low and featureless, that it seems, from one of the heights of Mendip or the Quantocks, to be one unbroken plain. On the western edge of this great level rise in turn the Quantocks, the Brendon Hills, and the broad upland of Exmoor, the most elevated district in the county.

A glance at the map will show that most of the marked physical features of Somerset, its hill-ranges and the courses of its rivers, lie in converging lines, like roughly drawn radii of a great circle whose centre is in the north-west of Wales, in the direction of the island of Anglesea. These lines may be regarded as creases in a gigantic "fold" which begins at the Belgian coal-field, and reaches to County Wicklow.

The chief ranges of hills are Exmoor, the Brendon Hills, and the Quantocks in the west, and the Mendips in the north of the county. Minor stretches of hill are Dundry, Broadfield Down and Lansdown—which may be regarded as north-easterly spurs of Mendip—and the Polden Hills, half-way between Mendip and Exmoor. There is also a good deal of hilly country along the eastern and southern borders, the Black Down range, in the far south-west, being the most important. And the numerous small isolated knolls that here and there break the otherwise dead level, such as Nyland, Lyatt Hill, Brent Knoll and Glastonbury Tor, are well-marked features of the great alluvial plain.

The great upland of Exmoor—named after the river Exe, which rises at Simonsbath, in the heart of it—lies partly in the county of Devon, but most of its area and all its highest points—of which the loftiest, Dunkery Beacon, is 1708 feet above the sea—are in Somerset. A space of about 20,000 acres in the heart of the moor, surrounded by a ring fence 52 miles in circumference, is known as Exmoor Forest. But it is scantily timbered, and there is no reason to suppose that it was ever covered with trees. The word forest, which in mediaeval times meant an open hunting-ground, as distinguished from an enclosed park, is here applied as it is in Scotland, where the deer forests, as a rule, have no timber at all.

The higher ground of Exmoor, especially about the skirts of Dunkery, on whose summit may be seen the stone hearths for the beacon-fires that in old times were lighted there, to give warning to the country round of the approach of danger, is a piece of wild and desolate moorland, over which the wayfarer may wander for hours without seeing a living thing except birds, or a few wild red deer or half-wild ponies, and without hearing a sound except the "wind in the heath," or the call of a curlew or the croak of a raven. But its varying colours, changing with the changing seasons,—its sheets of purple heather, its thickets of golden gorse, the vivid green of its swampy hollows, the sunburnt tinting of its wide grassy slopes, the sober greyness of its winter landscape, together with its solitude and remoteness and even its very silence, combine to render it, to the artist and the lover of nature, one of the most attractive spots in Somerset. Some of its many valleys, such as those of the Exe and the Barle, and the beautiful glen called Horner Combe, are well-wooded. Exmoor is a well-watered country ; almost every valley has its stream, and in this respect it differs widely from the Mendips, where so much of the water runs through underground and unseen channels. Scattered over its lower reaches are many picturesque villages, such as Winsford and Luccombe. And on its northern side,

Horner Combe and Dunkery

where it slopes steeply towards the sea, near Dunster and Minehead and Porlock, is some of the finest scenery in the west. In the extreme north-west is the Doone country—Oare village and Badgeworthy, and other spots made famous by the novelist Blackmore, who represents them as the haunt, in the seventeenth century, of a community of brigands who were the terror of the district.

Exmoor is noted for its breed of sturdy little ponies; and it is, moreover, the only place in all England where red deer have run wild from time immemorial. In the autumn, the chase of the deer, a sport which is followed in the same way nowhere else in the world, attracts crowds of visitors from all parts of the country.

The Quantocks, which are almost joined to Exmoor by the outlying spur of the Brendon Hills, but which form a range remarkably definite and compact, are much less extensive, and much less wild and picturesque in character; and their highest point, Will's Neck,—that is to say, Weallas Nek, or the Welshmen's Pass, the pass over which the Britons were driven by their Saxon conquerors—is only 1261 feet above the sea. But the hills are finely timbered on their western slopes, and their lines are broken by beautiful glens, such as Hunter's Combe and the Valley of the Seven Wells. Here also there are wild red deer, but they are of comparatively recent introduction, having crossed over from Exmoor within the last hundred years.

A special point of interest about the Quantocks lies in their associations with Wordsworth and Coleridge. The former wrote some of his *Lyrical Ballads*, and the latter some of his finest poetry, including especially the *Rime of the Ancient Mariner*, among these hills.

The Mendip Hills stretch right across the north of the county from the sea—where they divide into the three headlands of Brean Down, Worlebury, and Sand Point—

Alfoxton, Wordsworth's House

nearly to Frome, a distance of rather more than 30 miles. Seen from a distance, and especially from the south, where there are no outlying spurs, they have, for the most part, a flat and monotonous outline; and although some parts of their northern slopes are well-wooded, most of their southern side is singularly bare and treeless. There

Toplady's Cleft, Burrington Combe

is no point in them which rises head and shoulders above the rest of the range. The rounded mass of Black Down, their highest summit, is only 1067 feet high, while the bolder and much more conspicuous crest of Crook's Peak, which is used as a steering-mark by sailors in the Bristol Channel, is no more than 628 feet above the level of the sea. But there is among the Mendips much of quiet country charm. There are many beautiful spots in them, such as Winscombe Valley, Burrington Combe, Ebbor rocks and Vallis Vale; while the Gorge of Cheddar, whose magnificent cliffs tower to a height of 442 feet above the road that winds below, has no rival in the kingdom. The limestone of the Mendips is pierced by many caves; some, like those of Cheddar, famous for the beauty of their stalactites; others, like those of Wookey and Banwell, for the bones of extinct animals which have been found in them; and others again, like the Lamb's Lair and the Eastwater Swallet, for their length and intricacy, combined with some real beauty.

Perhaps the most characteristic feature of these hills is to be found in the district known as "On Mendip," an expression applied to the high ground east of Shipham and to the parts round Charterhouse and Priddy, which, although mostly cut up into fields and crossed by several roads, and once the seat of a great mining industry, is very thinly inhabited, and has an air of extreme loneliness and desolation. The historic associations of the Mendips, together with their camps, tumuli, and other antiquarian features, are of very great interest.

The broad flat country in the heart of Somerset, which

Cheddar Gorge

in its different parts is known by various suggestive names, such as Sedgemoor, the Huntspill Level, and the Brue Level, and has an area of between two and three hundred square miles, possesses a most distinctive character. A great part of it is below the level of the sea at high water, and would be constantly flooded at every spring tide were it not for the sea-walls and sand-hills along the coast. In 1607, when the wall near Burnham gave way, a tract of country 20 miles long and five miles wide, extending as far inland as Glastonbury, was flooded ten or twelve feet deep, with disastrous results to life and property, as we learn from a chap-book of the time.

The fields in the moor are divided one from the other not by hedges, but by ditches full of water, known in the district as "rhines," which serve not only as boundaries and for drainage, but to some extent as waterways for flat-bottomed boats. Very many of them are bordered with willow-trees, which are pollarded to supply osiers for basket-making; and these trees, with their broad heads, their grey and furrowed bark, and the bright colour of their young branches, are a most picturesque and striking feature of the moor.

With a view to prevent and regulate floods, a body called the Somerset Drainage Commissioners was established in 1877, not only to keep the embankments in repair, and to build and maintain pumping-stations for carrying away surplus water from the rhines into the rivers, but, by means of sluice-gates, to arrange for the systematic flooding of the land,—a process of great value and importance to the farmers.

This part is preeminently the country of the Seo-mere-saetan, the "dwellers by the sea-lakes," whose

A Somerset "Rhine"

Anglo-Saxon name has been modified into Somerset. The draining of the moor was gradually accomplished, and extended over a very long period of years. Meare

Pool, the last vestige of the inland sea, was drained about the year 1800.

In the names of many places in the moor is evidence that they were once surrounded by water, since they end in some modification of the Saxon *ea*, an island. Such, for example, are Athelney, Othery, and Muchelney. A considerable tract of ground a few miles from Bridgwater, slightly raised above the general level, is called the Zoyland, that is to say, Zealand, or the island.

An important industry of these moors is the cutting of peat, which is much used in the district for fuel. Peat is a mingled growth of various plants—grass and sedge and heather, and especially sphagnum or bog-moss,—which has taken ages to accumulate, and in some places has formed beds 18 or 20 feet deep.

Turf Moor is to the naturalist a most fascinating region, very rich in plants and insects and fresh-water shells. The almost innumerable rhines are haunted by solitude-loving birds, by water-shrews, and even by otters. The abandoned turbaries, the places where peat has been cut, are carpeted in summer and autumn with flowers of many hues. The great Osmunda or royal-fern flourishes in the swampy copses. In the ditches grow such plants as the graceful arrow-head, the delicate water-violet and the curious little bladder-wort. Here, too, by the tool of the turf-cutter, have from time to time been brought to light ancient canoes, primitive weapons, Roman coins and other relics, and even ornaments and implements of gold. The marsh-village of Godney in Meare parish, the most important remains of the prehistoric Iron Age ever found in this country, is described in some detail in another chapter.

5. Watershed and Rivers.

The rivers of Somerset are small, and, with the exception of the Bristol Avon, a short part of whose course belongs to this county, are of slight commercial importance. Most of them are, moreover, muddy and slow moving. The main watershed is well-marked and simple in character, lying along almost the entire eastern and southern border, in the hilly ground that parts Somerset from Wiltshire, Dorset, and Devon, sometimes on one side, sometimes on the other side of the frontier, and with a single branch or offshoot in the case of the Mendip Hills.

The best known of the Somerset rivers is the Bristol or Lower Avon, which, rising in Wiltshire, flows for a short distance through the north-eastern corner of this county, nearly encircling the town of Bath, to which point it is navigable for small vessels. It then forms part of the dividing-line between Somerset and Gloucestershire, passing, on its way to the Channel, through the great river port of Bristol. Below Bristol it flows through the deep and picturesque Avon gorge, which is crossed by the Clifton Suspension Bridge, a light and graceful structure, 245 feet above the river, erected in 1864, largely from the materials of a bridge that had formerly spanned the Thames at Hungerford. On the left bank, that is to say, on the Somerset side, are the beautiful Leigh Woods, sloping steeply to the shore. Although the shipping of the port of Bristol has for many centuries passed up the Avon, it is a narrow and difficult river, and the new docks at Avonmouth, on the Gloucestershire side of its entrance, opened in 1908 by King Edward, are of very great advantage to the city.

The most important Somerset river, as far as the shire itself is concerned, is the Parrett, which, with the aid of its tributary streams and of the network of rhines or ditches connected with it, drains the great central plain in the heart of the county. Rising near South Perrott, a mile inside the border of Dorset, it flows northward, past North Perrott and South Petherton, passing, at the latter village, a house called King Ina's palace, no Saxon building however, but dating from the reign of Richard II. Just before reaching Muchelney, famous for its ruined monastery, the Parrett is joined on its left bank by the Ile, a stream which gives its name to Ilminster, Ilton, Isle Abbots and Isle Brewers. The last word is a corruption of de Bruyère, the name of a former owner of the manor. After receiving on its right bank the waters of the Yeo or Ivel, from which Yeovil and Ilchester take their names, the Parrett winds round the historic little town of Langport, originally Llongborth, that is to say, "Ship-harbour."

Five miles below Langport the Parrett is joined, on the left bank, by its most important tributary, the Tone, which, rising in the Brendon Hills, flows at first nearly due south until it reaches the border of Devon, then turning east, passing through the county town of Taunton, and watering the fertile plain of Taunton Dean, it meets the Parrett about a mile north of the famous Isle of Athelney. Not far below the meeting-place of the rivers, on the right, close to the village of Weston Zoyland, is the site of the Battle of Sedgemoor. On the opposite bank, and once much nearer to the stream, whose course

The Bore ascending the Parrett

has altered in the lapse of time, is another place taking its name from the river Tone, North Petherton. The Parrett then flows through Bridgwater, the most important port in the county, widening out below the town, and forming at high tide a considerable stream. Here, just before high water, and particularly at spring-tides, is seen a phenomenon not uncommon in some tidal rivers and especially on the Severn—the "bore," a crested wave sometimes nearly three feet high, which rushes up the river, suddenly filling it to the brim, and changing it, in a few minutes, from an insignificant ditch to a broad river. Cromwell and Fairfax, reconnoitring the fortifications of

The Bore at Bridgwater

Bridgwater from an open boat, before they commenced to attack the town, were caught by the bore, and their little craft was nearly upset. Between Bridgwater and the sea the Parrett receives on its right bank an important tributary, called in the early part of its course the river Cary, and later, where it has been straightened and deepened so as to carry off more easily the surplus rainfall and the flood-water of the moor, known as King's Sedge Drain and the Sedgemoor Cut.

Another stream playing an important part in the drainage of the central plain is the river Brue, which rises near the modern structure called Alfred's Tower on the borders of Wiltshire, flows through one of the flattest

Sedgemoor in Flood

parts of the county, including Godney Moor, Burtle Heath, and the Brue Level, all of which are below high-water mark, and lends its name to South Brewham, North Brewham, Bruton, and Burnham. Two miles above the mouth of the Brue is the river-port of Highbridge.

All these moorland streams are fenced with high

earth embankments so as to guard against floods, which have, at various times, done great damage in this part of Somerset.

The Frome rises a few miles north of Bruton, and after passing the ancient town of Frome near the eastern extremity of the Mendip Hills, and forming part of the boundary between Somerset and Wiltshire, joins the Avon at Freshford. Another Frome, unconnected with this county, and wholly on the Gloucestershire side, flows into the Avon lower down, in Bristol itself.

In the middle of the town of Chard, which stands on the actual watershed, rise two streams, the Ile and one of the tributaries of the Axe, whose head-waters are also in this county. The former flows north to join the Parrett and the estuary of the Severn, while the latter runs south, through Dorset and Devon, to the English Channel.

The Exe and its tributary the Barle, which rise in the west of Exmoor, not far apart, meeting finally near Dulverton, are two fine streams, famous both for their scenery and their trout-fishing. The Exe crosses the border and goes south through Devonshire, and, by way of Exeter and Exmouth, to the English Channel.

Another Axe, wholly a Somerset river, runs out of Wookey Hole, near Wells, while its main tributary, the Cheddar Water, flows out of another cave in Cheddar Gorge. But the head-waters of both streams are on the top of Mendip, where both have, in former times, been polluted by the lead-workings, with disastrous consequences to the fish in the rivers below. The Axe is a small stream, but it is of historic interest, since it formed for a time the boundary between the Saxons and the Britons of West Wales. Until 1802 it was navigable for barges as high as Axbridge, which, however, is now a mile from its banks. At Uphill, at the mouth of the river, is a tidal harbour for small vessels. There was also formerly another little port on the Axe, at Reckley,

Wookey Hole

anciently Redcliff, near Compton Bishop, to which lead was brought down from the Mendip mines. On the north side of the Mendips is another Yeo, flowing out of the fine artificial lake called Blagdon reservoir. At Chewton Mendip is the source of the Chew, a beautiful little stream which, after passing Chew Stoke and Chew

Magna, near which are the stone circles of Stanton Drew, joins the Avon at Keynsham.

Other small rivers are the sluggish little Kenn, which drains Kenn Moor near Clevedon, and the picturesque streams which fall into the Bristol Channel at Watchet, Dunster, and Porlock. The Cale, rising in Pen Selwood, and flowing through Wincanton (originally Wyndcaleton) is a tributary of the Stour, one of the rivers of Dorset.

There are now no lakes in Somerset. The whole county contains no more than about 25,000 acres of inland water, and this is chiefly in the form of the ditches already alluded to. Meare Pool, a sheet of water 500 acres in extent, the last vestige of the inland sea that gave its name to Somerset, was situated three miles north-west of Glastonbury. It was drained at the beginning of the last century.

6. Geology and Soil.

By Geology we mean the study of the rocks, and we must at the outset explain that the term *rock* is used by the geologist without any reference to the hardness or compactness of the material to which the name is applied; thus he speaks of loose sand as a rock equally with a hard substance like granite.

Rocks are of two kinds, (1) those laid down mostly under water, (2) those due to the action of fire.

The first kind may be compared to sheets of paper one over the other. These sheets are called *beds*, and such beds are usually formed of sand (often containing pebbles), mud or clay, and limestone, or mixtures of these materials. They are laid down as flat or nearly flat sheets, but may afterwards be tilted as the result of movement of the earth's crust, just as one may tilt sheets of paper, folding them into arches and troughs, by pressing them at either end. Again, we may find the tops of the folds so produced washed away as the result of the wearing action of rivers, glaciers, and sea-waves upon them, as one might cut off the tops of the folds of the paper with a pair of shears. This has happened with the ancient beds forming parts of the earth's crust, and we therefore often find them tilted, with the upper parts removed.

The other kinds of rocks are known as igneous rocks, which have been melted under the action of heat and become solid on cooling. When in the molten state they have been poured out at the surface as the lava of volcanoes, or have been forced into other rocks and cooled in the cracks and other places of weakness. Much material is also thrown out of volcanoes as volcanic ash and dust, and is piled up on the sides of the volcano. Such ashy material may be arranged in beds, so that it partakes to some extent of the qualities of the two great rock groups.

The production of beds is of great importance to geologists, for by means of these beds we can classify the rocks according to age. If we take two sheets of paper, and lay one on the top of the other on a table, the upper one has been laid down after the other. Similarly with two beds, the upper is also the newer, and the newer will remain on the top after earth-movements, save in very exceptional cases which need not be regarded by us here, and for general purposes we may regard any bed or set of beds resting on any other in our own country as being the newer bed or set.

The movements which affect beds may occur at different times. One set of beds may be laid down flat, then thrown into folds by movement, the tops of the beds worn off, and another set of beds laid down upon the worn surface of the older beds, the edges of which will abut against the oldest of the new set of flatly deposited beds, which latter may in turn undergo disturbance and renewal of their upper portions.

Again, after the formation of the beds many changes may occur in them. They may become hardened, pebble-beds being changed into conglomerates, sands into sand-stones, muds and clays into mudstones and shales, soft deposits of lime into limestone, and loose volcanic ashes into exceedingly hard rocks. They may also become cracked, and the cracks are often very regular, running in two directions at right angles one to the other. Such cracks are known as *joints*, and the joints are very important in affecting the physical geography of a district. Then, as the result of great pressure applied sideways, the rocks may be so changed that they can be split into thin slabs, which usually, though not necessarily, split along planes standing at high angles to the horizontal. Rocks affected in this way are known as *slates*.

If we could flatten out all the beds of England, and arrange them one over the other and bore a shaft through them, we should see them on the sides of the shaft, the newest appearing at the top and the oldest at the bottom, as shown in the figure. Such a shaft would have a depth of between 10,000 and 20,000 feet. The strata beds are divided into three great groups called Primary or Palaeozoic, Secondary or Mesozoic, and Tertiary or Cainozoic, and the lowest Primary rocks are the oldest rocks of Britain, which form as it were the foundation stones on which the other rocks rest. These may be spoken of as the Pre-cambrian rocks. The three great groups are divided into minor divisions known as systems. The names of these systems are arranged in order in the figure with a very rough indication of their relative importance, though the divisions above the Eocene are made too thick, as otherwise they would hardly show in the figure. On the right hand side, the general characters of the rocks of each system are stated.

With these preliminary remarks we may now proceed to a brief account of the geology of the county.

The geology of the British Islands is perhaps as complicated as that of any other country in the world; and in no other county in England is there a greater variety of rocks than may be found in Somerset, where almost every kind of English geological formation is represented. It is this variety which has produced such wide differences both of natural features and of soil, from the picturesque scenery and comparatively sterile uplands of the older rocks—which upheaval and shrinkage of the earth's crust have raised far above their original position—to the monotonous but fertile levels of the much more recent great alluvial plain.

	Names of Systems	Characters of Rocks
TERTIARY	Recent & Pleistocene	sands, superficial deposits
	Pliocene	
	Eocene	clays and sands chiefly
SECONDARY	Cretaceous	chalk at top sandstones, mud and clays below
	Jurassic	shales, sandstones and oolitic limestones
	Triassic	red sandstones and marls, gypsum and salt
PRIMARY	Permian	red sandstones & magnesian limestone
	Carboniferous	sandstones, shales and coals at top sandstones in middle limestone and shales below
	Devonian	red sandstones, shales, slates and limestones
	Silurian	sandstones and shales thin limestones
	Ordovician	shales, slates, sandstones and thin limestones
	Cambrian	slates and sandstones
	Pre-Cambrian	sandstones, slates and volcanic rocks

There are two well-marked divisions in the geology of the county. Thus, while the geological features of the west of Somerset, from the shore of the river Parrett to the border of Devon, are simple in character, and the kinds of rock few in number, conditions in the eastern districts are altogether different. There, the geological system is complex in the extreme. Nowhere else in our island are there so many varieties of formation in so small an area, as occur in the east of Somerset.

The older rocks are seen at the two opposite ends of the county, in the west and the north-east, while the wide intervening space is filled with newer formations.

The great mass of Exmoor, the Brendon Hills, and the Quantocks, are almost entirely made up of very old rocks, Devonian and Upper Silurian, including slates, grits and sandstones, originally formed of mud and sand and deposited at the bottom of an open sea. In the oldest of these rocks, the Morte Slates, are found trilobites, and other fossil remains of very ancient forms of life. The Devonian formation contains iron ore—that in the Brendon Hills is of high quality—and it is also much quarried for building-stone and slate. The soil over it is mostly poor; and a great part of the uplands is covered with heather and bilberry plants, with oak trees lower down. There is evidence of volcanic action at several points among the Devonian strata, as, for example, at Hestercombe on the eastern skirts of the Quantocks, and at Withiel Florey, on the south slope of the Brendon Hills.

Another very ancient rock, though considered more

recent than the Devonian, is the Old Red Sandstone, which forms the core of the Mendip Hills, coming to the surface at Black Down and other points on the highest part of the range, and also at Sidcot Hill and the Church Knoll at Winscombe. It also occurs at Portishead, and at Abbot's Leigh in the gorge of the Avon. It is believed to have been deposited in an inland sea whose waters contained iron, which has tinged the rock with red. No fossils have been found in it in the Mendip country, but there are remains of fish in the Portishead beds. The Old Red Sandstone is not of much commercial value, and the soil over it is poor. Many of the Mendip springs rise at points where the limestone shales rest on this formation. For a long distance on the high ground of Mendip, between Downhead and Beacon Hill, there is an outcrop of the igneous rock called andesite.

Above the Old Red Sandstone, in order of time, comes the greatly more important Carboniferous or Mountain Limestone, which underlies the coal measures, and was originally from ten to thirteen thousand feet in thickness. The lowest layer of the series is the Lower Limestone Shale, a comparatively soft and crumbling rock containing many fossils,—trilobites, teeth of fish, and various bivalve shells such as spirifers. This bed is well seen in Burrington Combe, at the base of Callow cliffs, and in the gorge of the Avon. Upon the shales rests the Carboniferous or Mountain Limestone proper, a rock almost entirely composed of corals and other animal remains, and probably formed in clear deep water. It includes all the picturesque features of Mendip scenery;—

Cheddar Cliffs, the gorges of Ebbor and Burrington, Vallis Vale, the cliffs of Callow, the bold crest of Crook's peak, the rocky headlands of the coast near Weston-super-Mare, and the two Holms out in the Channel, as well as the beautiful combe of Brockley, in Broadfield Down. Characteristic of this formation are its many caves and combes, which have been formed by the dissolving away of the rock by the action of rain-water, and by subterranean streams, usually acting along lines of fracture.

There are more or less famous caves at Cheddar, Burrington, Wookey, Loxton, Uphill, Compton Bishop, East Harptree, and to the north of Wells, some of which, particularly those of Cheddar, and to a less degree the great cavern called the Eastwater Swallet, contain very beautiful stalactites, stalagmites, and draperies of stone, or rather of carbonate of lime, which is composed of lime dissolved out of the limestone rock by rain-water charged with carbonic acid.

A stalactite is an icicle-shaped formation hanging down from the roof of a cave, and is formed by dripping water, of which each drop, as it falls, leaves behind a very minute quantity of carbonate of lime. Some of the rest of the solid matter in each drop is deposited on the floor below; and as, in the lapse of ages, the stalactite slowly grows longer and longer, the stalagmite as slowly rises to meet it. In some cases the two have joined, forming a beautiful pillar, generally intensely hard, and often with a crystalline structure. So small is the quantity of carbonate of lime left by each falling drop, and so slow is the resulting formation, that there is at least

Gough's Stalactite Cave, Cheddar

one case to be seen at Cheddar in which a stalactite and a stalagmite are united by a single, though continually changing drop, and have for fifty years shown no further sign of uniting.

Several of the Mendip caves have been robbed of their beauty, but the stalactites of the Cheddar caves have been carefully preserved, and are believed to be the most beautiful in the kingdom. In some of the caverns have been found the bones of many animals now extinct, as will be noticed in more detail in the following chapter. In a fissure near Wells the geologist Moore found 70,000 teeth of one species of fish, as well as a number of teeth of a small kangaroo, the oldest mammal yet known to science.

A "combe" in limestone hills is in general a later stage of a cave whose roof has fallen in. It was in this way that the magnificent gorge of Cheddar was formed, and not, as was formerly supposed, by a violent convulsion of nature which tore the rocks asunder. The lines of the strata on both sides of the gorge are continuous, and lie at exactly the same slope, which would not be the case if the rocks had been rent asunder; and the sides of the ravine are, moreover, distinctly water-worn.

It may be added that the scarcity of visible streams is a feature of the Mendip country. The water finds its way almost entirely underground; partly, no doubt, through natural fissures, and partly through caves and galleries which it has hollowed for itself.

The original arrangement of the Carboniferous Limestone has been so altered by a great subterranean upheaval that there are places where the rock is completely folded over and turned upside down. There is a good example of this by the side of the road between

Vertical Limestone Strata, Churchill

Churchill Batch and Dolbury, where the strata are quite vertical, gradually sloping away in both directions. This formation contains a great variety of fossils, including corals of many kinds, bivalve shells such as *Spirifera*, *Productus*, and *Terebratula*, together with the univalve, *Euomphalus*, a few trilobites, and many encrinite stems. It is probably from this rock, which is there at a great depth below the surface, that the hot springs of Bath originally rise.

Several metals—lead, zinc, iron, and manganese—occur in the Carboniferous Limestone. These are described in another chapter. The stone itself is much used for building, and is also burnt into lime, both for making mortar and for agricultural purposes. The soil on the Mendips and their offshoots is thin, and is mainly devoted to pasture. There is an isolated outcrop of this formation at Cannington, not far from Bridgwater.

The Coal Measures of Somerset, which are the most southerly in England, have been so much disturbed by upheaval, and are so bent and folded, that at some points they now lie underneath the Mountain Limestone which they originally covered; and there are cases in which the same seam is passed through three times in the same shaft. In addition to that part of the Bristol coal-field which extends into this county, there are three basins in the Somerset coal-fields, of which Radstock is the most important, and Clapton-in-Gordano the smallest. That at Nailsea is no longer worked. It is probable that there is a large coal-field to the south of the Mendips, between Brent Knoll and Glastonbury, at a depth of perhaps 1000 feet.

Coal is made up of the remains of various plants, especially of giant club-mosses, horse-tails, and ferns, which grew in swamps near the mouths of rivers. Many fossil plants are found in the Somerset beds, particularly at Radstock.

After the Carboniferous Period the district suffered much change. Not only was it submerged beneath the sea, but 6000 feet or more of its solid rock appears to have been washed or otherwise worn away, to be re-deposited as mud and sand, which, in the lapse of ages, were converted into other formations.

After a long interval of time there were then deposited upon the edges of the upturned and denuded strata the New Red Sandstone or Triassic rocks, formed, it is thought, like the Old Red Sandstone, at the bottom of an inland sea containing iron, which has coloured much of the stone red; a sea liable to inroads of salt water, and surrounded by drifting sands. The various beds of this series form a rough half-circle in the east, south-east, and south of the county, from Dundry to the Mendips, past Wells, Cheddar, and Axbridge, skirting the southern side of the Polden Hills, round the Quantocks, past Minehead and Porlock, round Taunton and Wellington, and along the bases of the Blackdown Hills. One of the beds, dolomitic conglomerate, an ancient sea-beach made up of fragments of older rocks cemented together, skirts parts of the Mendips and Broadfield Down. It is an excellent building stone, and is much quarried at Draycott.

The New Red Sandstone contains some iron ore, a good deal of lead and zinc, which, however, are no longer worked, and a valuable ore of strontium, called, from its blue colour, celestite. Potato-stones are also characteristic of it. They are round nodules, sometimes hollow and lined with beautiful crystals. In the Avon gorge remains of great fossil lizards have been found in this rock. The New Red Sandstone soils are among the most fertile in Somerset. The rich vale of Taunton Dean is a good example.

The Rhaetic or Penarth beds of the Trias, whose thin strata overlie the New Red Sandstone, and are better developed in this county than anywhere else in England, are well seen at Watchet. They contain many striking fossils, and the teeth before mentioned as having been found in a fissure in the limestone had come from a Rhaetic bed. Cotham "landscape stone," which owes its tree-like markings to the presence of manganese, also belongs to this formation.

The Lias, which was largely formed of the mud and clay resulting from the wearing away of earlier limestone rocks and was deposited in a shallow sea, occupies a large part of the flat country in the middle of Somerset, together with considerable areas near Bath and Bristol. Some of the conspicuous outlying knolls, such as Glastonbury Tor and Brent Knoll, are partly formed of this rock, fragments of which have, in these cases, survived, while the rest of the bed of which they once formed part has been washed away.

The Lower Lias is noted for its fossils, of which the most numerous are ammonites, while the most striking are the remains of gigantic lizards,—plesiosaurus and ichthyosaurus, of which many fine examples have been obtained from quarries at Street. Lias is much worked for building-stone and, to a still greater extent, is burnt for making cement and lime. Great slabs of it, set on end, are much used for fencing.

The Midford Sands, which connect the Lias and the

Plesiosaurus, from Street

Oolite, and yield the fine brown Ham Hill building-stone, also cap the lias of Brent Knoll and Glastonbury Tor.

The Oolites, both Inferior and Great, which contain many fossils, form broad bands down the east side of the county, and both provide excellent building-stone. The best, in the former case, is worked at Doulting. A variety not quite so good, but particularly rich in fossils, is quarried at Dundry. The Great or Bath Oolite, although less durable than either Ham Hill or Doulting Stone, is more familiar than either, under the name of freestone. The city of Bath is practically all built of it, and it is used all over England. Between the two kinds of Oolite lies the fuller's earth, well developed at Crewkerne. It is a clayey bed, and is largely used for brick-making, as well as for cleansing purposes. Down the east and south-east of the county run the Cornbrash and the Oxford Clay.

Cretaceous formations are very sparingly represented in Somerset. Gault and Upper Greensand are found near Chard, in the Blackdown Hills, and in the extreme east. The Greensand yields material for scythe-stones, which are much exported. Chalk, elsewhere a most important formation, and one which formerly extended over all Somerset, now occupies only a small area between Chard and Crewkerne.

The broad levels of the basins of the Parrett and the Brue, and the flat country between Yatton and the sea, are in great part covered with a deep alluvial deposit of clay and sand and gravel, left by the once much broader waters of the Bristol Channel, and the once far greater stream of the river Severn. Above this have been formed deep beds of peat, much dug for fuel in the turf moor district, described in more detail in a former chapter.

7. Natural History.

The ancestors of the present land-animals of this country, its quadrupeds and its reptiles, its land-shells, and even the majority of its birds and insects, must, like most of its trees and plants, have reached it during the time when what we now call the British Isles formed part of the mainland of Europe. That that condition was, however, of comparatively short duration is shown by the fact that this colonisation was not complete; that there are fewer kinds of wild animals in England than there are in Germany, for example, and that there are fewer still in Ireland than there are in England. Long isolation always brings about differentiation of species, and hence it is clear that it is not so very long, as geological periods go, since Great Britain again became an island; for it possesses no quadruped or reptile, and only one bird, the red grouse, which is not also to be found in Europe. Very different is the case in Japan, which was separated from Asia so long ago that forms of life have had time to vary, so that the islands abound in species which are unknown on the adjacent continent.

Many creatures, once inhabitants of this country, have died out, either because the climate changed and so cut off their food-supply, or because they were destroyed by man. Of some of these the skeletons still remain, and Somerset is more rich than any other part of England in relics of such creatures as the mammoth, rhinoceros, tiger, lion, bear, hyaena, and wild horse; animals which once roamed over the hills of the West country, and whose bones have been discovered in the Mendip caverns.

Although there are more kinds of beasts and birds on the mainland of Europe than there are in this country, both birds and beasts are much more abundant here. Nothing strikes a naturalist more forcibly, when travelling on the Continent, particularly in France or Italy, than the scarcity of wild life, especially of birds. We have fewer species, but many more individuals. To this, various causes have contributed. We do not, as is the custom in European countries, shoot and trap for food small birds of every description. And game-preserving, although it has been fatal to the larger birds of prey, such as kites, falcons, harriers, and buzzards, and keeps down other species, such as jays, magpies, and carrion-crows, provides innumerable sanctuaries for great numbers of small birds, where they are safe from harm during the breeding-season.

The natural features of Somerset are so varied in character, including as they do the hill-country of Mendip, the wilder and less cultivated heights of Exmoor, the low-lying levels of the turf-moors with their countless water-ways, and the long stretch of sea-shore, that the county is very rich in both animal and vegetable life. Very nearly every land-quadruped native to Britain is found, or has been found in it, from the tiny little pygmy shrew, of which it takes ten to weigh an ounce, to the tall red deer, which, in this country, is only found wild on Exmoor and, of recent times, upon the Quantocks. Badgers are not uncommon in many of the wooded districts; otters are regularly hunted on the Exmoor rivers and elsewhere, and in some places foxes are still only too plentiful. The polecat is now rare, and the

Head of Red Deer

marten still more so. But stoats and weasels, squirrels, rats and mice, all the voles, shrews, both land and water, moles, several kinds of bats, hedgehogs and rabbits are all abundant. Hares have become less so of recent years. Several kinds of marine mammals have been observed in the Bristol Channel; dolphins, grampuses, and porpoises, for example. Both the common seal and the harp seal have been killed on the coast, and a live Greenland whale was once stranded at Weston-super-Mare.

One of the charms of the Somerset landscape lies in the abundance and variety of its birds, the list of which, both of residents and regular visitors, is a very long one. Of rarer species, the raven and the buzzard are familiar objects on Exmoor; the former is frequently seen among the Mendips, while the peregrine falcon still has an eyrie on Steep Holm. Another characteristic Exmoor bird is the blackcock, which, of recent years, has also established itself among the Mendips. As in other parts of England, the hawfinch and the woodcock are here extending their breeding-range, and are more common now than formerly, and the nightingale, although here nearly at its western limit, is not uncommon. Great numbers of waders, such as sandpipers, plovers, curlews, and oyster-catchers, together with many kinds of gulls and great flocks of ducks, especially scaup ducks, frequent the sea-shore in winter, and a few gulls breed on Steep Holm. But razor-bills, guillemots, puffins, and cormorants are comparatively rare stragglers. There are several heronries and innumerable rookeries, while the multitudinous jackdaws that harbour in Cheddar Cliffs are a feature of that beautiful gorge. It was in this county that the first black stork, the first Egyptian vulture, and the first hawk-owl ever seen in Britain were recorded. The few kinds of reptiles and batrachians present no points of special interest beyond the fact that the first palmated newt recorded in this country was found near Bridgwater in 1844.

The variety of marine fish recorded for the Somerset coast is considerable, but the actual number of each species, owing to the muddiness of the water, is small. The larger kinds that have been taken include sword-fish of eight feet in length and sturgeon of ten feet, with halibut of 40, sea-angler of 80, and sun-fish of 180 pounds' weight. Flying-fish have been taken at Burnham. Some of the more important kinds, from the fisherman's point of view, are mentioned in another chapter. Salmon are less common than they were, but the great reservoir at Blagdon provides some of the best trout-fishing in our island.

Few if any other counties of England have so many kinds of land and fresh-water shells. A little spiral species, called, for want of an English name, *Bulimus montanus*, very rare in most parts of the country, is common among the limestone screes of Mendip, and there is abundance of another kind, so small that it takes 210 dead specimens of it to weigh a single grain. The marine life of Somerset, in spite of the great length of the coast-line, is scanty and uninteresting, owing partly to the muddiness and partly to the comparative freshness of the water.

All the different kinds of British butterflies except five—and some of these five have been reported—have been taken in Somerset, including even undoubted examples of the Apollo and the Large Copper. The

Bath White was named from a piece of needlework executed at Bath from a specimen of the insect said to have been caught near that city, about the year 1795.

Somerset is a county of flowers. Even from the

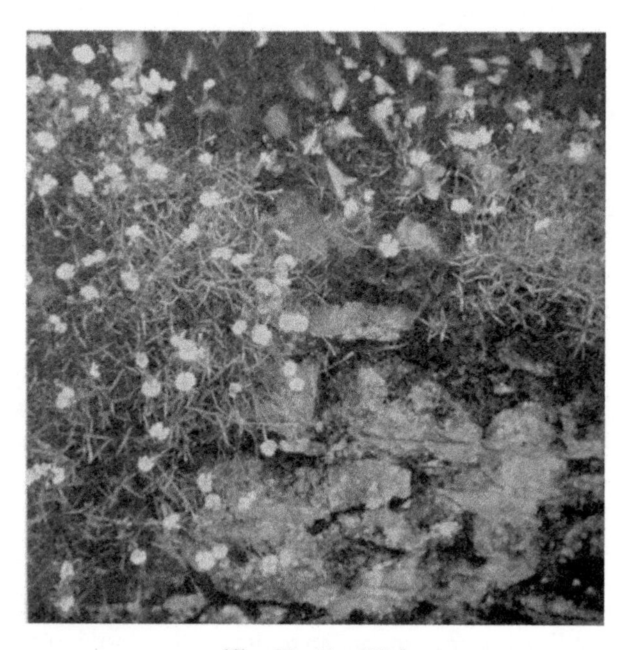

The Cheddar Pink

railway-lines the traveller may realise something of its beauty, while a closer acquaintance with its hills and woods and meadows, its sea-shore and its water-ways, reveals an extraordinary amount of interest and charm. The flora varies not only with the altitude but with the geological formations. The plants that love the Old Red Sandstone differ in marked degree from those that flourish on the Carboniferous Limestone. The foxglove, for example, grows on the former but not on the latter. The hill-plants, again, are widely different from those both of the turf-moors and of the coast. Not only are there in the county a great many kinds of flowering plants, about 1050 all told, most of them beautiful, and some of them very rare, but there are five species, among which are the Cheddar Pink (*Dianthus caesius*) and the Steep Holm Paeony (*Paeonia corallina*), which are found nowhere else in Britain. And there is only one other habitat in the island for the delicate little rock-rose (*Helianthemum polifolium*), which is so abundant on Brean Down as sometimes to whiten the southern slope of that pleasant promontory like a light sprinkling of snow. Lichens are very abundant on the trees of Somerset woods, an evidence, it has been suggested, of the purity of the air.

Trees of many kinds grow well in Somerset, especially in the west, where may be seen some of the noblest English examples of the Spanish chestnut, cypress, oriental plane, sweet bay, walnut, and Luccombe oak—a well-marked variety which originated in the county. The finest walnut in England is said to be the magnificent tree growing at Cothelstone, and the finest oriental plane that at Lydeard St Lawrence. The elm, which is generally believed to have been introduced into Britain by the Romans, is perhaps the most characteristic of Somerset trees, but it is a tree of the open field and of the hedgerow rather than of the woodland. At Somerset Court there is an avenue of elms 1000 yards long. The ash is also very widely distributed, and is probably equally common. It is especially a tree of the limestone, and is very abundant along the Mendips. It has been suggested that it was probably the principal tree in the ancient Mendip Forest.

Holford Beeches

The Ashen Fagot Ball, a festival long kept up at Taunton, was believed to commemorate that winter night when the soldiers of King Alfred found, to their joy, that green ash branches made excellent camp-fires. The beech grows well in Somerset, especially in the west, although it is almost certainly not indigenous. Extensive oak-woods are found in the east and south-west, and there are magnificent oaks in some of the Exmoor valleys and in the neighbourhood of Dunster. There has been much afforestation in Somerset since about 1750. Birch and Scotch fir have been a good deal planted on Turf Moor.

In addition to many names of places which are based on names of trees, such as Ashwick, Martock, and Bicknoller, called respectively after the ash, oak, and beech, we have parishes of Oak, Ash, Elm, Alder, Thorn, and Hazel.

8. A Peregrination of the Coast.

The coast-line of Somerset, which measures rather more than 60 miles in length, is for the greater part of that distance, that is to say, from the mouth of the Avon nearly to Watchet, with slight exception flat and featureless, and much of it below the level of the sea at high tide. On the skirts of Exmoor there is, however, some fine scenery, the shore near Porlock Weir and Minehead rising in bold and precipitous wooded steeps, but there are practically no cliffs on the coast of Somerset. Those at Watchet, although very picturesque, are low and insignificant, and the bolder steeps of Brean Down are of no great altitude. A very marked character of the whole coast is the great distance to which the tide goes out, leaving exposed, not only fine sands, but vast banks of mud. And at many points the currents are swift, and dangerous to bathers.

Rather more than two miles west of the mouth of the Avon is the pleasantly situated little seaside town of

Portishead, a popular watering-place, and with a harbour sheltered by a low, wood-covered hill, on whose slopes the houses of the town are built. The most interesting point about the place is its very high tides, which usually rise 42 feet, and at the equinox 45 feet. They are the highest in the Old World, and are only exceeded in the

Coleridge's Cottage, Clevedon

Bay of Fundy, between Nova Scotia and New Brunswick, where the rise is 50 feet.

Six miles further down the coast is Clevedon, a very pretty and much frequented watering-place, although it has no sandy beach such as forms one of the chief attractions of Weston-super-Mare. It has interesting associations with Coleridge, and Tennyson's friend

Hallam was buried in its parish church. Clevedon Court is one of the most remarkable old residences in England. An ancient encampment, one of the three Somerset Cadburys, stands on a hill near the town.

Four miles beyond Clevedon is the mouth of the Yeo, really an insignificant little stream, but with a considerable estuary at high tide. A low rocky headland running out into the sea, due west of the mouth of the Yeo, is known as Sand Point or St Thomas's Head. The first name is taken from Sand Bay, closely adjoining it, and the second from the ruined priory, partly dedicated to Thomas a Becket, which stands at the foot of its southern slope. On its northern side may be traced the remains of a raised sea-beach, consisting mainly of sand and pebbles, firmly cemented together by the lime of the abundant limpet and periwinkle shells. A fine stretch of sand joins this headland to the next, which is known as Worlebury. Its heights are crowned by one of the most remarkable hill-forts in Somerset, and at its foot lies the little island of Birnbeck, now united to the mainland by a pier, where steamers call, and where there is a life-boat station. Sprat-fishing, once a most important industry here, is now of little value. In the cliff opposite Birnbeck has been found another raised sea-beach, in which could be seen ripple-marks, rain-drops and the footprints of birds, as well as bones and teeth of the horse, ox and other animals.

Along the south side of Worlebury, and covering a large space of the low-lying land at the foot of it, is the attractive and popular modern watering-place of Weston-super-Mare, famous for its mild climate which has made it an important health-resort, for its many attractions to the holiday-maker, and for its magnificent sands, more than two miles long, extending to the mouth of the river Axe.

At the far end of Weston beach is the last spur of the main range of the Mendip Hills, with the ruined church

Weston-super-Mare

(General view from Worlebury Encampment)

of St Nicholas on the top of it, and the little port or Uphill, noted for its antiquarian and geological interests, at its base. On the other side of the estuary of the Axe, in which rises the Black Rock—anciently an important boundary mark, and useful to sailors because it is never covered by the tide—is the promontory of Brean Down, which forms a good shelter for the port of Uphill, and is a most attractive spot to the naturalist and the antiquary; a habitat for rare plants and a haunt of interesting birds. At the end of the down is an abandoned fort, in which were formerly mounted seven heavy guns.

Out in the Channel lie two small islands, Flat Holm, five and a half miles from Weston-super-Mare, and Steep Holm, three miles from the extremity of Brean Down. On the former, which now belongs to Wales, are a light-house, an isolation hospital, and a crematorium in connection with the port of Cardiff. In 1867 a raised sea-beach was discovered on this island, in which were not only shells, but horns of red deer, and even part of a human skull. On Steep Holm are the scanty remains of a very ancient priory, founded it is thought by the Berkeleys, but of which very little is known. The wild paeony, for which this is the only English habitat, may have been introduced by the monkish residents. On both islands there were formerly batteries and small garrisons.

Beyond Brean Down is another splendid stretch of sand, bordered by a ragged line of sand-hills, or "tots," as they are called in the district, and extending to the mouth of the Parrett, a distance of seven miles.

The marine life of this shore is poor, but the plants and molluscs of the sand-hills are of much interest to the naturalist. Wreckage of various kinds is often washed in under the southern shore of Brean Down, and the cottagers who live near the beach easily collect from it sufficient drift-wood for their fires. Among the jetsam are not infrequently found buck-eye beans that have come all the way from the West Indies. In hot weather a mirage is sometimes seen here, representing, even when the tide is far out, a phantom sea, in which houses and other details of the shore are clearly reflected. Close to the sea, two miles and five miles respectively from Brean Down, are the villages of Brean and Berrow; and two miles further still is Burnham, another very popular watering-place, near the mouth of the Brue—from which it takes its name—with some shipping, two lighthouses, and a life-boat station. The roadstead is protected by the island of Steart.

At the mouth of the Brue begins the broad estuary of the Parrett, the most important river of Somerset. For several miles beyond this point, the coast which so far has been, with slight exceptions, absolutely flat, still continues very low, but it rises as we approach Watchet, and becomes more and more picturesque as we go further west.

Watchet is one of the best of the small Somerset tidal harbours, a life-boat station, a place of some historic interest, especially in connection with Saxon times, and of considerable natural beauty. The low but very picturesquely coloured cliffs near the little town are of red conglomerate, in some places overhanging the sea, and curiously veined with alabaster of varying hues, from white to brilliant red.

Two miles beyond Watchet is Blue Anchor, a little cluster of houses named, as is the case with several west-country hamlets, from its inn. The features of this part of the coast are the pebbly beach, with sand outside it, and the low, flat, shelving rocks that fringe the shore. For some distance beyond Blue Anchor the railway, which runs close to high-water mark and has frequently been damaged by the waves, is protected by baulks of timber.

Minehead

From Blue Anchor to Minehead the coast becomes flat again, with no definite line between the shingle of the beach and the grass of the low-lying fields.

A long spit of sand, running far out at low tide, helps to shelter the snugly-situated harbour of Minehead, which lies mainly at the foot of a bold headland called the North

Hill. Minehead, a place of some historic interest, was once an important sea-port, with considerable trade to the Mediterranean, and a favourite landing-place from Ireland. But it is now visited only by excursion steamers and small coasters, and is more familiar as a watering-place. It is still better known as the Gate of Exmoor, and as a great resort in the stag-hunting season. At several points along this part of the coast, especially off Stolford, Minehead, Blue Anchor, and Porlock, there may be seen on the beach, at low tide, stumps of the trees of a submerged forest, amongst which have been found not only mammoth tusks and other animal remains, but rude implements of flint.

Beyond Minehead the coast rises abruptly into precipitous wooded steeps, sheer up from the sea, and extending to Hurlstone Point, a fine rocky headland, with a coastguard signalling station on the top of it. Between Hurlstone Point and Gore Point is a broad stretch of flat marshy land which has been reclaimed from the sea, and is defended by a huge bank of shingle. On the innermost edge of this flat expanse is Porlock. Its name means "the enclosed harbour," but it is now a mile or more above high-water mark. The present harbour, Porlock Weir, visited by a few small coasters, is more than two miles away.

Beyond Porlock Weir the coast rises once more, reaching at one point a height of 1200 feet, and forming a continuous wooded steep, broken by a series of glens. In one of these, on a little plateau about 400 feet above the sea, is the tiny but picturesque little village of Culbone, whose church, 35 feet long, is one of the smallest in all England. The county ends in the valley of Glenthorne, the house of that name being on the Devonshire side of the border.

Culbone Church

9. Coastal Gains and Losses. Sand-banks. Lighthouses

At some points along the coast of Somerset the sea has gained upon the land, as is shown by the remains of ancient forests which may be seen at very low tides near Minehead and Porlock, and Stolford. Near Weston-super-Mare, on the other hand, decayed fishing-stakes, an anchor, and a primitive canoe have been found at spots far distant from the present tide-line, showing that the land has gained upon the sea. These changes took place at a period very remote, but since man was in possession of the country. In more recent times much land near the coast, formerly frequently covered by the tide, has been reclaimed and brought under cultivation, and the sea kept back by walls which some think were built by the Romans, and by heaps of sand which are held in place by marram-grass and other kinds of vegetation. So large a tract of ground has been reclaimed from the sea in Porlock Bay that the town of Porlock, once a port, is now separated from the water by more than a mile of meadow-land, as has been already stated.

On Bridgwater Bay the sea is now gaining on the shore. A large part of Steart Island, between the mouth of the Brue and the town of Burnham, has been washed away during the last hundred years, together with a farm-house and out-buildings which once stood at the north-west corner of it. Not more than sixty years have passed since a house on the shore opposite the island, the road that formerly led out to Steart, and a ridge of pebbles called Chesil Beach, have all been swept away by the sea. In the entrance of the river Parrett, again, is a small island on which were once corn-fields, and a notorious inn, which claimed to be outside the jurisdiction of the law, and on which no dues were paid. Not only has the inn been washed away, but much of the island itself, which now only affords pasturage for a few sheep. On the other hand, there is a piece of ground on the right bank of the Brue, near Highbridge, which, only half a century ago, was part of the river bed.

As is so often the case with the estuaries of great rivers, the Bristol Channel is much obstructed by banks of mud and sand, which, partly because of the enormous quantity of solid matter continually being brought down by the Severn, and partly because of the strong tidal streams, are continually changing in shape and position, thus adding greatly to the difficulties of navigation.

Liable to storms, intricate in its navigation, and destitute of any harbour of refuge, the Bristol Channel is one of the most dangerous parts of the British seas. In the last complete year for which returns have been made out by the Board of Trade, there were, round the coasts of the United Kingdom, 3685 accidents to ships, including 280 total wrecks, and involving a loss of 269 lives. About one-tenth of these, that is to say, 371 accidents, 27 total wrecks, with a loss of 25 lives, were recorded for the Bristol Channel, reckoning from Hartland Point to St David's Head. Most of the disasters happen off the Welsh shore. There is comparatively little sea-traffic on the Somerset side. The fairway for Cardiff, Bristol, Avonmouth and even Portishead, lies on the Welsh side of Steep Holm.

The most important obstacles to navigation on the coast of Somerset, in addition to the general shallowness of the water, are the banks known as the English Grounds, extending three miles out from Clevedon, and the Gore Sand, Graham Bank, and Culver Sand in

Bridgwater Bay. The last of these, a shoal nearly five miles long, is covered by water varying in depth from a quarter-fathom to two-and-a-half-fathoms. The less important Longford Grounds, a bank lying inside the English Grounds, may possibly preserve a Norse name for the Bristol Channel—the Long Fiord.

To warn the sailor against these and other dangers, buoys, lights, and bells are provided at various points; but these, again, are not numerous on the Somerset coast. There are minor lights connected with the piers at Portishead, Clevedon, Weston-super-Mare, Watchet, and Minehead, and there is an "unwatched light" on Blacknore Head, near Portishead. But the one solitary important lighthouse is that at Burnham, which is at a considerable distance from the sea, and which, from its pleasant situation and surroundings, is regarded as affording one of the most comfortable posts in the service. The tower of Burnham lighthouse is 99 feet high, and the lantern shows a 10,750-candle power white light, of the group-flashing order, with two eclipses in the minute, and visible for 15 miles. Five hundred yards nearer the tide-line is another and lower tower, only 36 feet high, with fixed red and white lights. The Flat Holm lighthouse, which is one of the most conspicuous and most familiar off this coast, belongs to Cardiff. The tower, which in 1737 was built to a height of 69 feet, was raised to 99 feet in 1820. The group-flashing lantern, with four eclipses to the minute, gives a white light of 35,000 and a red light of 14,000 candle-power visible for 18 miles. A fog-siren of great power has also lately been established here.

Most of the English lighthouses are in the charge of the Trinity House, a corporation which was founded in 1512. It has now a revenue of £300,000 a year, derived from "light-dues" levied on shipping. The early beacons were fires of coal. One of these was in use at St Bees Head as late as 1822. There are now round the coasts of Britain more than 1000 lighthouses and lights of various sorts and sizes, from those of the first order, such as those at the Lizard and at St Catherine's Point, which are the most brilliant in the world, and may be reckoned in millions of candles, down to tiny little structures, of which there are many, like the 100 candle-power "Jack-in-the-Box" on the river Tees.

The Somerset coast is provided with four life-boats of the best modern design, stationed at Weston-super-Mare, Burnham, Watchet, and Minehead.

10. Climate and Rainfall.

The climate of any country, or, in other words, its average weather, by which we mean its temperature, rainfall, and hours of sunshine, as well as the dryness or otherwise of its air, depends upon various circumstances and conditions, but especially upon the geographical position of the country, that is to say, upon its nearness to the equator, upon its distance from the sea, and upon its elevation; partly also upon its soil and vegetation. Speaking generally, the nearer we approach the equator the hotter will be the climate, and the nearer we are to the sea-coast the milder and more equable will it be. It is not, however, the fact that the greatest heat is found near the equator, nor the most intense cold in the immediate neighbourhood of the poles. The highest temperature ever recorded in the world, 127° Fahrenheit, was in the Algerian Sahara, at a spot not even in the tropics: and the greatest degree of cold ever experienced, 90° below zero, Fahrenheit, was at a place in Siberia, only just inside the Arctic Circle.

The climate of the British Isles is very greatly influenced by the ocean-current, often called the Gulf Stream, from its supposed source in the Gulf of Mexico, which generated and impelled by steady winds carries a constant stream of warm water from the equatorial region towards the North Pole. This current washes the shores of these islands, and it is that circumstance which makes our winters so much milder than those in Labrador, which is no nearer to the pole than we are. But for it our weather would be worse than that of Newfoundland.

The highest temperature ever experienced in England was 101°, at Alton in Hampshire, in July, 1881; and the lowest was 10° below zero, at Buxton, in February, 1895. A temperature of 23° below zero, Fahrenheit, is said to have been measured in Berwickshire, in 1879, but the correctness of the record has been questioned. What is of more importance, however, is the average temperature, which for England alone is 48°. The south is, of course, the warmest part of the country, but the east is both colder and drier than the west.

Somerset, like Devon and Cornwall, has a mild and

equable climate, though naturally to a less degree than either of those counties, which are further west and further south. The winters are seldom severe. Snow is not common, and rarely lies long upon the ground. At Weston-super-Mare geraniums and calceolarias have been known to survive through the winter, out-of-doors, for three years in succession; while round Dunster such shrubs as the myrtle and camellia flourish freely in the open air, and some palms grow well without protection.

The mean annual temperature for the whole county is not known, but for Weston-super-Mare it is $52\frac{1}{2}°$, or more than $4°$ above the average for all England.

The sunniest districts of England are, as might be expected, on the shore of the English Channel, though the southern parts of both the east and the west coasts also get more sunshine than inland places in the same latitude. The sun is above the horizon all over Britain for more than 4400 hours in a year; but, owing to the frequent presence of clouds, he is not visible for even half that time in any part of the island. The most favoured districts—the southern edges of the counties on the English Channel, and the shores of Kent, Essex, and Norfolk—get no more on an average than 1800 hours of bright sunshine in the twelve months, while the manufacturing centres in the interior, owing to the smoke which so often obscures the sky, get only 1200 hours or even less, which means about three hours a day all the year round. The sunniest months throughout the country generally are May and June, and the gloomiest month is December. The average duration of sunshine in Somerset is between 1500 and 1600 hours in a year, or rather more than four hours a day. In 1906, which was an exceptionally sunny year, the county enjoyed from 1800 to 1900 hours of sunshine, or an average of about five hours a day all the year round; although the actual amount would, of course, be much more than this in the summer and much less in the winter.

The prosperity of any particular country and the pleasantness of its climate depend not only on the sunshine but on the rain. The amount of rain that falls in any place depends in its turn chiefly on the height of that place above the sea, on its distance from the western coasts, and on the configuration of the ground, that is to say, upon its position with regard not only to hills or mountains, but with respect to valleys, up which moisture-laden air may be driven by the wind, to be compressed and cooled until its moisture falls in rain. When we speak of an inch of rain we mean that it would lie an inch deep on a perfectly level piece of ground. An inch of rain all over an acre of ground weighs rather more than 100 tons.

The rainfall varies very much in different parts of England; but the average amount for the whole country is about 33 inches in a year. That is to say, if all the rain that fell in a year stayed on the ground, and did not evaporate or run away or sink into the earth, the water would be 33 inches deep all over the country at the end of the twelve months. Not only does the rainfall vary in different parts of the country, but it varies in different years. The wettest year on record was 1903, when the rainfall averaged 50 inches for all England, while the

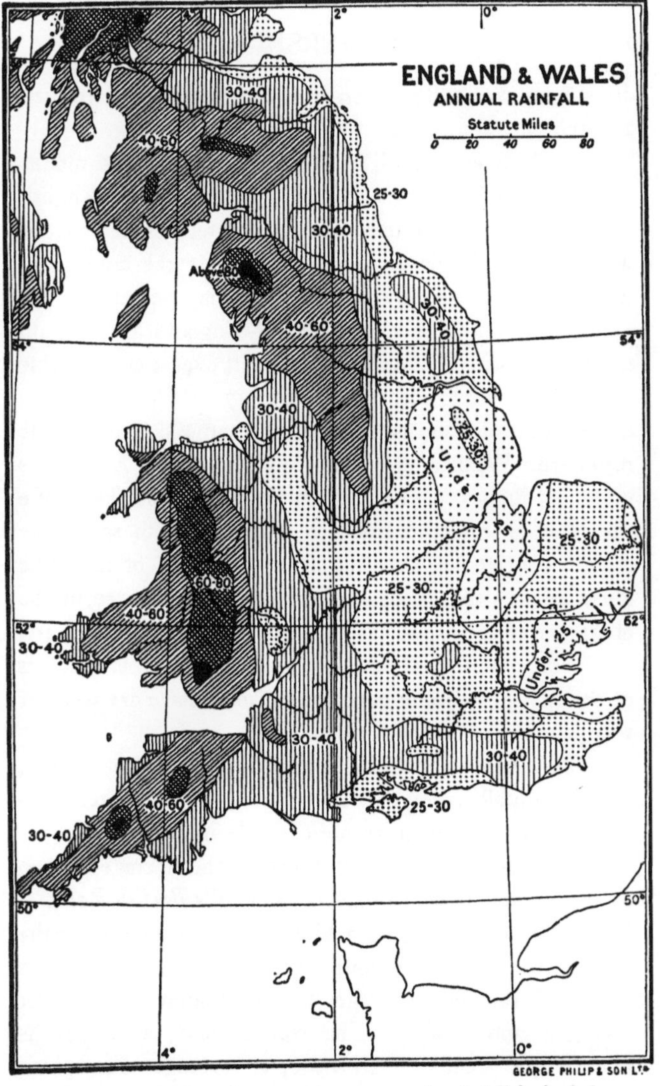

(The figures give the approximate annual rainfall in inches)

driest year was 1887, when it amounted to no more than 24 inches.

The heaviest rainfall in these islands is in the mountains—round Ben Nevis and Snowdon, and in the English Lake district, and it may fairly accurately be said that as we go across England from west to east the amount of the annual fall steadily decreases, as may be seen in the annexed map. The driest parts of England are thus on the east coast. While the average fall over a considerable part of the west side of the country is from 40 to 60 inches, that on the east is no more than 25 to 30; while round the Wash and in part of Suffolk and Essex it is under 25 inches. Thus it sometimes happens that while the west of England has rain enough and to spare, the eastern districts are suffering from the want of it. The effect of this difference is shown in some degree in the character of the crops. The farms of the rainier west are, to a great extent, laid down in grass. The drier districts of the east grow more corn, which must have dry weather to ripen it.

The driest month in England generally is March, whose rainfall averages 1·46 inches; and the wettest is October, in which the amount is 2·84 inches.

The rainfall for Somerset, taking the average of 100 of the stations named in Dr H. R. Mill's *British Rainfall*, is slightly over 34 inches, which is only an inch above the average for the whole country, and the number of days on which rain falls averages 186, compared with 181 for England and Wales. The wettest part, as might be expected, is Exmoor. In 1906, an average year, 70·49 inches of rain fell at Simonsbath, which is 1080 feet above the sea. In the same year 47·8 inches fell at Chewton Mendip, on the north slope of the Mendip Hills, while the low-lying districts averaged about 25 inches.

Sea-fogs not unfrequently blow in from the Bristol Channel, where they are often very heavy in the winter, usually when the barometer is high; and land-fogs are common in the low ground south of the Mendip Hills. The former may come on at any hour, but especially towards nightfall. The latter, which often hang over the marshes and even on the hill-tops in the early mornings, generally disappear as the day advances. There are two sayings about them in the Mendip country:—

> Fog on the hill brings grist to the mill,

and

> Fog on the moor brings sunshine to the door.

11. People—Race, Dialect, Settlements, Population.

The earliest inhabitants of Somerset of whom we have any knowledge, the people of the Palaeolithic or Early Stone Age, have left few traces beyond their weapons and implements of flint. They were hunters, and appear to have practised no craft but that of hunting, while their arts seem to have been almost if not entirely limited to the use of fire and the making of rude instruments of stone.

But during the Neolithic Period, as the later Stone Age is called, this district was invaded by an Iberian race from south-western Europe, a race possessing flocks and herds, with a knowledge of many arts and crafts, such as spinning and weaving, the making of pottery, and of dug-out canoes, but having, at first, no knowledge of the use of metal. They were of the same stock as the Silures of South Wales, probably dark-haired and black-eyed, round-headed and short of stature, though authorities differ on some of these points. Their descendants may, it is considered, still be seen in the county, especially near Pen Selwood; and their breeds of domestic animals are believed to be still represented upon Somerset farms.

The Iberians were, it is thought, conquered and driven westward by the very different Goidels or Gaels, a powerful Celtic race of tall, fair-haired, long-headed men, much further advanced in arts and crafts, users of bronze for tools and ornaments, builders of the stone circles of Stanton Drew, and makers, in all probability, of many of our hill-forts, although it is considered that Cadbury, near Clevedon, is the only one which can be certainly attributed to them. They left their mark in the names of some of our hills and rivers, Dundry and the Axe, for example. Many of the people of Scotland and Ireland are their descendants, and their language is still spoken in the Highlands and Western Isles of Scotland, in parts of Ireland, and in the Isle of Man.

The Goidels were in their turn succeeded and conquered, in the fourth century before Christ, by the Brythons, another Celtic race, who gave their name to the island. They took possession of Wales, and of Scotland as far as the Highlands, but they do not appear to have crossed into Ireland. They were, to a great extent, users of bronze, but they also worked in iron, and were the first of the Iron Age in this country. It is probable that they built most of the camps in Somerset, and that they made many of the roads, some of which were afterwards adapted and improved by the Romans. The recently discovered lake-villages near Glastonbury and Meare were two of their settlements. It has been said that they have left more traces in Somerset than in any other county in England.

Shortly before the landing of Julius Caesar, Britain was invaded by still another Celtic race, the Belgae from Gaul, a tall, dark-haired people, as may be gathered from the appearance of their descendants, the Walloons of Liege and the Ardennes. It is believed that at the time of the Roman conquest the eastern half of Somerset was occupied by the Belgae, while the western half, and part of the north coast, was inhabited by the Brythons, Goidels and Iberians.

It is thought that the Roman occupation of Somerset, although it lasted for more than three centuries, had no great effect upon the character of the inhabitants. But, in the three successive waves of Saxon conquest, such of the Celtic population as were not destroyed were driven partly to the comparatively barren uplands of Mendip, and partly to the western parts of the county, especially to the Quantocks and Exmoor.

It is very probable that in the first Saxon invasion under the pagan Ceawlin, the original inhabitants of east Somerset were to a great extent destroyed. But his successors Kenwalch and Kentwine were nominally Christians, and they would therefore be less ruthless in their treatment of the vanquished. We see from the Laws of Ina that many Britons, probably in the west of the county, were spared, although chiefly, perhaps, reduced to a condition of slavery.

At the present day the inhabitants of the eastern half of Somerset are, as a rule, tall, with light hair and dark eyes. Those of the western half, beyond the river Parrett, which, for a long period, marked the limit of Saxon conquest, are shorter of stature, with darker skins, darker hair, and lighter eyes. The latter people are, in fact, more like the Irish, while the former are typical English. The people of east Somerset are Saxons, pure and simple, save for inevitable admixture; those of the west are Saxonised Britons. Again, in the hill-country of Mendip, of the Quantocks and Exmoor, and in the uplands of the south-east, not only do we find abundant evidence of the persistence of later Celtic or Belgic types, but also, in peculiar shapes of skulls—some very narrow, and some strangely pyramidal—vestiges, possibly, of the earlier races.

But what most of all distinguishes the inhabitants of the two regions is the difference in dialect. It is not so much a question of vocabulary, although there is considerable dissimilarity in that. The Britons of the west learnt the Saxon speech of their conquerors, but they pronounced it in their own way; and to this way they have adhered for more than 1200 years. The difference is so marked that a man from the east side of the county has a decided difficulty in understanding a man from the west side of the Parrett. The dialect of the east is much the same as that of Wiltshire, Hampshire, Berkshire. The dialect of the west is quite different, and resembles that of the county of Devon.

The following are a few out of many hundred peculiar words still current in the county, given by Dr Prior in the *Proceedings of the Somersetshire Archaeological Society*:—

axen, *ashes.*
bannut, *walnut.*
bivver, *to tremble.*
clamper, *a difficulty.*
drang, *a narrow lane.*
dumpsy, *towards twilight.*
dunch, *deaf.*
fess, *gay, smart.*
gout, *a gutter.*
gruff, *a mine-working.*
ligget, *a rag.*
lurden, *a sluggard.*
mang, *to mix.*
mooch, *to stroke.*
oseny, *to foretell.*
pilm, *dust.*
plim, *to swell.*
rexen, *rushes.*
scrint, *to scorch.*
shrig, *to lop a tree.*
sog, *wet ground.*
ted, *to turn hay.*
trig, *neat, to make neat.*
vang, *to catch.*
wevet, *a spider's web.*

One of many peculiarities is the sound of the diphthongs *oo* and *ou*, which are sounded like the French *u* or the German *ü*. Another is in the great variety of the vowel-sounds, and in the indistinctness or modification of the consonants. It is often said that a Somerset man turns every *s* into *z*, and every *f* into *v*. This is, however, only true with words of Saxon origin. Thus *seed* is changed into *zeed*, and *fan* into *van*. But *serpent* always keeps the sharp *s*, and *family* its sharp *f*.

Again, in the western districts, *th* and even *v* are often sounded like *dh*. A few of the many words peculiar to west Somerset and Devon are apparently genuine Norman survivals, such as *suant*, even, from suivant; *levy*, a flat place, from levée; and *payse*, to lift, from peser. It has been suggested that it may also be Norman influence which leads a Somerset man to say *bwoil* and *pwoint*, instead of boil and point. There were many Norman settlers in Somerset. Many double names like Huish Champflower, Wotton Courtenay, Shepton Mallet, suggest the addition of a Norman family title to the existing Saxon name of a manor. Huguenots and other French refugees have also at various times settled in the county. After the dissolution of the monasteries, the Duke of Somerset established a number of woollen-workers from the Low Countries in the domestic buildings that had belonged to Glastonbury Abbey; and at a later period Oliver Cromwell is said to have induced a number of Flemish weavers to settle in the same town. There can be little doubt that colonists such as these have handed down peculiarities of feature and manner which can still be traced, as well as such names, common especially in the Mendips, as Thiery, Sawtell, Cary, and Fountain.

It should be remembered that some of the forms of Somerset dialect which strike the educated ear as ungrammatical, are really survivals of pure Saxon speech, such as was in use at the courts of Alfred the Great and of Edmund Ironside. English in other parts of England has changed. In Somerset it has, in some respects, kept closer to its original form.

The population of the geographical county of Somerset,

according to the census of 1901, is 508,256, or, in round numbers, half a million; and there are in the county rather more than 100,000 inhabited houses. A hundred years ago the population was 273,577, or slightly more than half what it is at present. In the busier county of Kent the population has, in the same time, increased from 268,973 to 1,348,841. In common with all except five of the counties of England, there are more women than men in Somerset; the excess of the female over the male population being no less than 37,000. More than 31,000 people, or one-twelfth of the inhabitants, are engaged in agriculture, and about 7000 in mining or in quarrying stone.

Somerset is a somewhat thinly-populated county. There are in it about two acres to every man, woman, and child, or 320 persons to the square mile, compared with 558 to the square mile in the whole of England and Wales. Westmorland, the most sparsely inhabited county, has only 82 people to the square mile, or eight acres to each inhabitant. Lancashire, on the other hand, contains more than 2300 people to the square mile, or four people to every acre; and in the county of Middlesex there are 12,669 to the square mile, or about 20 inhabitants to every acre of ground.

12. Agriculture—Main Cultivations, Woodlands, Stock.

The area of all the land in England is in round numbers 32 millions of acres, of which 24½ millions are under cultivation, in addition to more than 2 million acres of mountain and heath land which may be used for grazing but which are not reckoned in the measurement of farms, and 1¾ million acres of woods and plantations. Of the 24½ millions of acres under cultivation, 10¾ millions are arable and devoted to crops, while 13¾ millions are down in permanent grass. For the last fifteen years the area of cultivation in the whole country has been gradually growing less; the decrease in England last year having been 15,000 acres, chiefly in wheat, barley, potatoes, turnips, and swedes. The cultivation of fruit, on the other hand, is steadily increasing, and amounts now to nearly 300,000 acres. With regard to live-stock, the government returns show that the total number of horses in England (about a million), and of cattle (about 5 millions), were less in 1907 than in 1906, but that the number of sheep (15 millions), and of pigs (2¼ millions), had considerably increased.

All counties in England are more or less agricultural, but Somerset is pre-eminently so, having few other commercial or industrial interests; and in its own particular and special kinds of grazing and dairy-farming it is surpassed by few shires in the country.

According to the latest return of the Ordnance Survey, Somerset contains rather more than a million acres (1,043,409), of which 854,931 acres are cultivated, including 675,964 acres laid down in permanent grass, which is four-fifths of the entire cultivated area. This amount of permanent grass is only exceeded in the much larger counties of Yorkshire and Devon, while the proportion of grassland to the total area devoted to crops is much greater than in either of those two counties. Corn crops, which include not only wheat, barley, oats, and rye, but peas and beans, occupy only one-tenth of the cultivated area, or 84,000 acres. In this respect Somerset is surpassed by 20 other shires, including the adjoining counties of Gloucester, Wiltshire, and Devon; while the space devoted to corn crops in Essex and Lincoln is one-third, in Norfolk two-fifths, and in Cambridge and Suffolk nearly one-half of the respective cultivated areas.

Green crops—potatoes, turnips, swedes, mangolds, rape, and vetches—occupy 44,000 acres, in which Somerset is sixteenth among the counties of England in the actual area devoted to such crops. It may be observed that the yield of mangolds per acre in Somerset (26 tons), is much higher than the average for the whole country (19 tons) and higher than in any other county in the island. The position of Somerset is about the same in sainfoin, clover, and grasses under rotation, which occupy 46,000 acres, the county being here fourteenth among the English shires. It is worth noting that the yield of hay to the acre of permanent grass in Somerset, in 1907 (25 cwts.), although surpassed by several other counties, was greater than the ten years' average for the whole country. Flax is a crop grown far more extensively in Ireland than in England, but is interesting to know that Somerset produces not only more flax than any other English county, but one-fifth of all that is raised in the country. Teasels, used for raising the nap of cloth, were formerly much grown in the parishes of Wrington, Blagdon, Ubley, Compton Martin and Harptree. When the

Somersetshire cloth industry declined they were exported to Yorkshire being packed on flat "dillies," like a hurdle on two wheels, and sent all the way by road. They have to a great extent gone out of cultivation since the

Teasels

introduction of artificial substitutes made of wire and leather, but there are some specially-faced cloths for which the substitutes do not answer, and teasels are still grown for the purpose between Taunton and Ilminster. Woad was formerly grown near Keynsham.

Somerset is not a county for small fruits. For example, while Kent grows one-third of all the strawberries that are raised in England, Somerset produces only one-sixty-fourth, a large proportion of which are grown on the south slope of the Mendips, especially near Axbridge and Cheddar. The apple-orchards of the county,

A Cider Orchard

however, are among the most extensive in the island, occupying one-fortieth of the whole of Somerset, or no less than 25,000 acres, an amount which is exceeded only in the much larger county of Devon, while it is double that found in Kent; and the proportion of orchard-land to the area of the whole shire is greater than in any other county in England. When orchards of other fruits, such as plums, pears, and cherries, are included in the total, Somerset falls behind a little, but even then is only surpassed by Hereford and Devon. Somerset apples are, however, largely grown for cider-making only, and much of such fruit is of no value for the kitchen or for dessert. Vines are grown against many Somerset cottage walls, as is the case in other parts of England; but the county formerly possessed several really important vineyards. At the time of the Domesday Survey there were vineyards at Glastonbury, Pamborough, and Meare, all of which belonged to the abbot, and there was a much larger one belonging to the crown at North Curry. At a spot near Bath, still called The Vineyards, Black Cluster and Muscadine vines, trained on stakes, were grown as late as 1730. In 1719 they produced 69 hogsheads of wine. At Claverton, again, a few miles further out of Bath, was a particularly flourishing vineyard, which, with others in the county, was in good bearing order at least as late as 1820.

Under the head of fruit-trees may be mentioned the walnut, which grows well in Somerset. There are many fine walnut-trees at Bossington, while what is said to be the finest in England is at Cothelstone.

Somerset once contained several large forests, but, although it has a good deal of scattered timber, we do not generally think of it as a specially well-wooded county. But its woods and plantations are considerable, and measure one-twentieth of its area, or nearly 47,000 acres, and it ranks twelfth in this respect among the shires of England. Sussex has the greatest proportion of woodland, amounting to one-seventh of its total area, and Cambridge the least, or only one-ninety-second. At the time of the Norman Conquest there were four "forests" in Somerset, —Mendip, Selwood, Exmoor, and North Petherton. A fifth, Neroche, was added later. But it is probable that Selwood and Petherton were the only two of these which contained many trees.

The principal kinds of trees have been mentioned in a previous chapter. It may be added that, in the low-lying districts, especially in the south-west, there are extensive withy-beds, as they are called, in which dwarf willow-trees are grown in the water in order to supply thin and pliable boughs for basket-making.

The total number of agricultural holdings of all sizes in Somerset in 1907 was about 14,000, which is more than in any other county except Devon, Lancashire, Lincoln, and Yorkshire. Of these holdings, 3367, or about one-fourth, measure less than five acres, and there are only two other counties, Lincolnshire and Yorkshire, which have more of these small farms.

As we should expect in a district so largely devoted to grazing and dairy-farming the numbers of the various kinds of live-stock in Somerset are large, and the county ranks high under these heads. In sheep (471,000) it stands ninth, in horses (41,000) it is seventh, in cattle (242,000) it is fifth, while in pigs (128,000) it is exceeded only by Suffolk and Yorkshire.

The most famous agricultural production of Somerset is Cheddar cheese, to which further allusion is made under the head of manufactures. Judging from government returns, it is the best thing of the kind made in all England.

13. Industries and Manufactures.

The county of Somerset does not possess a single manufacturing town of consequence, nor is it anywhere the seat of any really important industry apart from those that are connected either with agriculture or the raising of minerals, such as coal and building-stone.

The most famous production of Somerset is Cheddar cheese, which although named after a single village, is now made in large quantities, not only in the low-lying and fertile districts, especially in the east and centre of this county, but in the adjoining counties of Wiltshire and Dorset. Much "Cheddar cheese" of high quality is also made in America, by the aid of "starters," or laboratory-made preparations containing the necessary bacillus, called *Bacillus acidi lactici*, which is a microscopic organism shaped like a very diminutive figure-of-eight, and without which the cheese cannot possibly be made.

Cheddar cheese has been famous for centuries. John Locke, the philosopher, was a very successful maker of it. As far back as the reign of Henry II, more than 1000 pounds' weight of Somerset cheese, costing about a farthing a pound, was bought "to the use of John the King's son," as we learn from a paper in the Record Office.

Another characteristic product is cider, which is made in large quantities in almost all parts of Somerset, but especially in the western districts, where lie a large proportion of the county's 25,000 acres of apple-orchards. Cider-making is chiefly carried on at farm-houses, in

A Cider Press

many of which old and primitive appliances are employed. Modern methods, however, are rapidly coming into use, owing chiefly to the experiments of the official agricultural analyst for the county. Many kinds of apples are used, but the most characteristic are Kingston Black and some of the Jersey varieties. It is the opinion of experts that Somerset cider, of which ten million gallons are made in a good season, is the best in England. Were the area now given up to cider apples devoted to the growing of fruit fit for the table, the people of Somerset would gain in more ways than one.

The manufacture of Bath bricks, so much employed for cleaning knives and other objects made of steel, almost as famous as Cheddar cheese, and probably even more widely used, is entirely confined to one spot in Somerset, and is carried on nowhere else in the world. Bath bricks, which have no connection with the town of Bath, and are named after their inventor, are made at Bridgwater, from the mingled sand and mud of the tidal waters of the river Parrett. On both sides of this tawny stream, not far below the town, are a number of shallow pits or terraces, 10 or 12 feet broad and from 20 to 40 yards long, called "slime-batches," cut in the river bank. Into these, at slack tide, the water of the river is admitted through sluice-gates; and in them the solid matter in the water settles down, forming, in the course of a year, a dense stratum of a sort of clay, 12 feet thick. This is dug out in the winter and made into the familiar oblong blocks, which are baked in ordinary kilns.

An industry that has already been alluded to is the

cutting of peat, a great deal of which is dug in the moors near Glastonbury, the denser parts for fuel, and the lighter and more porous for making "moss-litter" for use in stables. Stacks of drying peats may be seen in great numbers from the railway that, joining Burnham and Templecombe, crosses the heart of the county.

Nailsea Glass
(*Taunton Castle Museum*)

A good deal of woollen cloth is made at Frome, Taunton, Twerton, and especially in the form of serge and blankets, at Wellington. But in the villages near the Mendips, where much cloth was formerly woven, the industry has long been discontinued. A thousand people are employed in the lace-works at Chard; and leather-glove-making provides work for a large part of the population of Yeovil and its neighbourhood.

Bricks, roofing-tiles, and drain-pipes are extensively made at Bridgwater, Highbridge, Crewkerne and elsewhere. Basket chairs are made at Curry, from osiers grown in the district. The bronze group of figures on the Thames Embankment, representing Queen Boadicea and her daughters, was cast at Frome, a town once famous for its bell-foundries. At the close of the eighteenth and the beginning of the nineteenth centuries glass was made at Nailsea and Bridgwater, and the conical red-brick furnace-houses are conspicuous objects at both places.

14. Mines and Minerals.

Somerset has been a mining district from time immemorial; but no metal is now worked here in any quantity, and the chief minerals that are raised in it at the present time are coal, building-stone, and clay.

Lead and zinc were formerly very extensively mined in the Mendips, where the Romans, who may have made use of the abundant ore of zinc, very early established lead-mining settlements, both at Charterhouse and Priddy, taking possession, no doubt, of mines that had been previously worked by the Britons. Operations were continued long after the Romans had withdrawn from the island; and the mining industry of the Mendips, including the working of both lead and zinc, had attained to such importance at the time of the Wars of the Roses that, when King Edward IV sent Lord Justice Choke to settle a dispute between the miners and the mining-lords, the assembly that met at Green Ore, near the highest point of the hills, is said to have amounted to 10,000.

A very curious code of mining-laws was drawn up on that occasion, from which it appears that, although a miner had to obtain a licence to dig for ore, the licence could not be refused, and that the miner might sink a shaft wherever he pleased within the mining forest, pro-

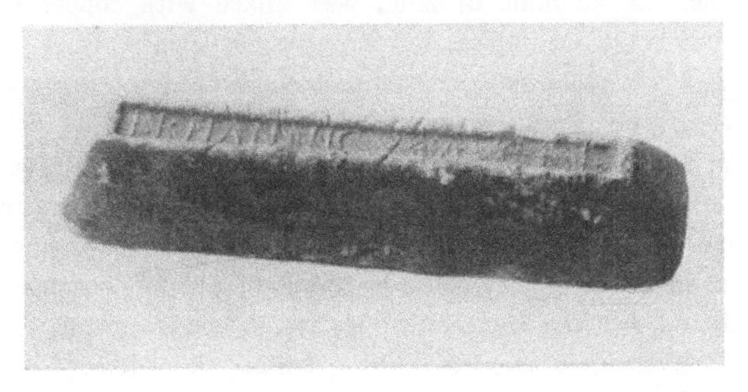

Roman Pig of Lead, cast A.D. 49, found near Blagdon
(British Museum)

vided that he paid one-tenth of the ore to the lord of the soil. One of the laws enacted that any man caught stealing ore to the value of thirteen-pence halfpenny, should be shut up in his house, with all his tools, and the house burnt over his head. He was apparently to be allowed a chance of escape, for the next law declared that if he was found stealing ore a second time he should "be tried by the Common Law, for this Custom and Law hath no more to do with him."

When the veins of lead were exhausted it was found profitable to re-smelt the refuse that had been left from the early and imperfect methods, and some of the slag was even treated twice over. Actual lead-mining has long ceased in the Mendips, but the re-working is still continued at Priddy, on a small and diminishing scale.

Zinc, which was obtained from two varieties of ore, calamine or carbonate of zinc, and blende or black-jack, otherwise sulphide of zinc, was mixed with copper to make brass. Shipham and Rowberrow, small villages on the Mendips, were great places for the calamine diggers, who, dressed in smocks stained deep red from the soil, worked in rough shallow shafts. The ore is not exhausted, but it is poor, and when the shafts became too deep to manage without machinery they were abandoned, since ore of better quality could be more cheaply obtained from abroad, and the Somerset mines are no longer worked.

There has been no mining among the Mendips for nearly 50 years, although several recent attempts have been made to search for iron. But there are wide areas among the hills, in what was known as the Mining Forest (which was not the same as the Hunting Forest) where the ground is so honeycombed by old shafts, and so broken up with the mounds and hollows left by ancient workings, as to be of little value except for grazing.

The Romans are believed to have extracted silver from the Mendip lead, and a large block or pig of lead was found, some years ago, at Charterhouse, marked EX.ARG.VE., which may possibly mean "from the silver-bearing vein." At a later period Somerset silver was used in the royal mint. Some of the coins of Charles II, William III, and the three following monarchs are stamped with a rose, which means that they were struck from silver which had come from the west of England, and probably from this county. Fifty years ago an attempt to mine silver was made in the Quantocks.

The Romans had iron-mines in the Brendon Hills. The ore, which is of high quality, is still worked, and there is a mineral railway from the workings down to the coast at Watchet. The iron-ore which occurs in the Mendips and on Broadfield Down is chiefly ground up into Spanish brown or yellow ochre, according to the nature of the material. A better quality of ore is raised and smelted at Ashton Vale, near Bristol. But the total amount of iron produced in the whole county is inconsiderable, and does not exceed 1000 tons in a year.

Copper is found in the Quantocks, but not in paying quantity, and manganese, which was formerly mined in several places, is no longer worked. A valuable ore of strontium called, from its blue colour, celestite, is obtained in the hills near Wells,—one of the two places in England from which this rare substance is exported to Germany for use in the extraction of sugar from molasses.

The coal-fields of Somerset are the most southerly and also among the thickest in Britain. Those that are worked lie in the north of the county, between Bristol and the Mendips, and consist of several distinct parts, the more important of which are the Radstock coal-field and the Bristol coal-field,—a part of which only is in Somerset. The seams of the former, although very numerous, are thin, and the coal is not of the best quality. Much better coal is obtained from the Ashton Vale collieries, just outside Bristol. The outlying coal-field at Nailsea, where some of the dismantled buildings are not unpicturesque objects from the railway-line, is worked out, and the pits have been abandoned for many years. The total amount of coal raised every year in Somerset is about a million tons.

The county is famous for several fine kinds of building stone, of which the most familiar, by name, is the popular Bath stone, or freestone,—so-called from the ease with which it is worked. A better and more durable material is Doulting stone. Pennant stone, formerly much used for pavements and for the coping of parapets, has been largely superseded by slabs of concrete. Slate is quarried in the west of the county. Five thousand tons of fuller's earth, and more than 200,000 tons of clay are raised every year. The latter is chiefly used for making bricks, roofing-tiles, and drain-pipes.

The mineral springs of Bath, which, from very ancient times, have been celebrated for their healing properties, are the hottest in the kingdom, attaining, in one instance, 120° Fahrenheit; much too warm for the hand to bear with comfort. The waters are especially beneficial in cases of gout, rheumatism, and paralysis, and have been described as the most important of their kind in the world. They are rich in lime, and in sulphates and chlorides, and they have recently been found to contain radium, which may possibly be the source of their curative powers.

15. Fisheries and Fishing-stations.

The sea-fisheries of England form one of her most important industries. Not only do they provide employment for 110,000 men and boys, including the crews of 9000 ships and boats, in the catching of the fish, and for an army of people engaged in distributing the 7½ million pounds' worth that are brought to our shores every year by British ships alone, but they furnish an immense quantity of cheap and wholesome food which, by rapid methods of transit, is available in all parts of the country.

The most productive of our fishing-grounds is the North Sea, and more fish are annually landed at Grimsby, Hull, Lowestoft, and Yarmouth than at all the other fishing-ports of England put together. Much fishing is also done by English trawlers off the shores of Iceland and the Faroës, and the boats now go as far even as the White Sea and the coast of Morocco.

About half the fish are taken by trawling, which consists in dragging a net, attached to a beam of wood, along the bottom of the sea, in comparatively shallow water. Very many kinds of fish are caught in this way, but haddock, plaice, and cod are by far the most numerous. Drift-nets and seine-nets, which are let down in the open sea, attached to floats of cork or to air-bladders, without regard to the depth, and sometimes a considerable distance from shore; and stake-nets, fastened to posts fixed in shallow water near the land, are also much used. Herrings are the chief fish caught in drift-nets and seines, and more of them are landed than of any other kind of fish. More than a million pounds' worth, each, of herrings, haddocks, and plaice, and more than three-quarters of a million pounds' worth of cod are brought into English fishing-ports every year. Pilchards, which are full-grown sardines, are caught in immense quantities off the coasts of Devon and Cornwall, and are chiefly salted and sent to the Mediterranean. Many fish, especially halibut, cod, and ling, are taken with hook and line, sometimes at great depths. Crabs and lobsters are mostly caught in wicker traps, and oysters usually by dredging.

Although Somerset is a maritime county, with a sea-board more than 60 miles in length, it has now no fisheries of real commercial importance. The rows of stakes that are still to be seen in many places along the shore are evidence that great numbers of nets were formerly set there. There was a time when the little port of Minehead sent 4000 barrels of herrings to the Mediterranean every year. But the herrings, as has happened in many other places round the British coasts, have almost entirely left these waters. Sprats, again, were once so abundant that £10,000 worth have been taken in one season between Burnham and Weston-super-Mare. But for many years they have not been plentiful, and though great quantities were caught off Steep Holm in 1908 the fishery is now usually of trifling value. A few herrings are still caught, and cod and whiting are fairly abundant. Oysters are still taken, though in much smaller numbers than formerly, at Porlock Weir. Large halibut have been caught off the Somerset shore, and sturgeon of 300 pounds' weight have been taken in the river Parrett.

The most widely distributed of important freshwater fish is the eel, which is to be found throughout the various ponds and water-ways in the county. Eels are less in demand for food than they once were. It is recorded in Domesday Book that the two fisheries belonging to Muchelney Abbey produced 6000 eels in a year.

It is only of late years that the life-history of this fish has been clearly understood. It is now known that eels, which are some years old before they are mature, go down in the autumn to the sea, off the western shores of the British Isles, and deposit their eggs at a depth of at least 6000 feet. The old eels never return to the rivers or lakes where they had spent their lives, and are believed to die in deep water after spawning. When the young fish are two years old they leave the sea and ascend rivers, making their way also to distant streams and ponds across fields and roads and even over stone walls. The young eels, known as elvers, come up the Parrett in vast numbers in the spring, and are caught in great quantities and made into cakes.

16. Shipping and Trade. Chief Ports.

The shallow waters of the Bristol Channel prevent the approach of large vessels to any point on the coast of Somerset. The county has no sea-ports worthy of the name, and such coastal harbours as there are can only be reached at high tide. The best of such harbours is Watchet, now well protected by a pier and breakwater. But only one vessel, on an average, enters it daily, while its chief import of coal amounts to no more than 18,000 tons in a year. Its chief export is flour, and since the

Watchet Harbour

re-opening of the Brendon mines there is an increasing shipment of iron-ore.

A fine dock, with large granaries, has lately been constructed at Portishead by the Corporation of Bristol, in order that large vessels may unload without going up the Avon.

The little harbour of Minehead, tidal although it is, is regarded as one of the safest on the Channel. It was formerly an important port, and was a favourite landing-place from Ireland, but its trade is now inconsiderable.

The Somerset town with most shipping is Bridgwater, a river-port on the Parrett, which is here navigable, at high water, for ships of as much as 900 tons' burden. The majority of the craft that visit the port are small

The Port of Bridgwater

colliers and coasters, but there is some foreign trade with Norway, Sweden, Russia, Germany, Holland, France, the Mediterranean and America. The chief imports are coal, timber, grain, raw hides, and valonea (acorn-cups for use in tanning); and the principal exports are building bricks and tiles, cement, and Bath bricks, of which eight millions are annually sent away from Bridgwater by sea alone.

Highbridge, on the Brue, is another small river-port, and Pill, on the Avon, is a pilot-station, with between 35 and 40 registered pilots.

The trifling importance of the Somerset ports may be gauged from the fact that at the last census only 40 sea-going and coasting ships, with 104 men and boys, are entered as belonging to the county, while the total tonnage of all the ships that enter Bridgwater in a year is only slightly over 120,000 tons, or about one-thousandth part of that entering the port of London in the same period.

17. History of Somerset.

The history of Somerset begins with the Roman occupation. For although we have many traces in the shape of hill-forts, burial-mounds, tools and weapons, pottery and ornaments, of the men who lived and died here before Julius Caesar landed in Britain, those people left no written records, and they are not mentioned by any ancient writer.

The Romans, who came to our shores in B.C. 55, were early attracted to Somerset partly by its mineral wealth, and partly by the hot and health-giving springs of Bath, or as they called it Aquae Sulis (or Solis). Relics of the conquerors are very abundant in the county, and they enable us to form a clear idea of the time when the district was occupied. They settled at Bath at least as early as 44 A.D., and the oldest known Roman inscriptions yet discovered anywhere in Britain are those which are stamped on two pigs of lead found near Charterhouse-on-Mendip which show that the invaders were working the lead mines of Somerset in 49 A.D. The latest of their coins, of which many thousands have been found in various parts of the county, bear lettering which proves them to have been struck late in the fourth century, that is to say, between 370 and 383 A.D.

The great West Saxon conqueror Ceawlin, after his victory over the Britons at Deorham, in 577, is believed to have established his frontier along the Mendip Hills (whose heights are crowned by a long line of camps or hill-forts, many of which, however, may have been made before his time), and the little river Axe, at the foot of the range, is considered to have marked the limit of his conquest. Kenwalch, a century later, carried the English border to the river Parrett, beyond which the country was called West Wales, and was still in the hands of the Britons. In 710, Ina, one of the ablest of the West Saxon kings, builder of the first Taunton Castle, founder of Wells Cathedral, and of a monastery at Glastonbury which was the fore-runner of the famous abbey, defeated the British king Geraint, and extended the English conquest still further west. After Ina's time there appears to have been no more fighting between the Britons and their invaders, and the fact that there were no fewer than eleven Somerset towns and villages at which Saxon coins were struck seems to suggest a condition of peace and even of complete subjection.

Somerset was closely associated with the history of the

Saxon kings, many of whom were in the habit of coming to hunt the red deer in the Royal Forest of Mendip. One of them, Edmund, son of Edward the Elder, nearly lost his life in the pursuit, his horse becoming unmanageable on the brink of Cheddar Cliffs, over which the deer had leaped. It was an incident that had a great effect upon the politics of England. In the moment of danger the king repented of the injustice with which he had been treating Dunstan. Not only was the young priest restored to favour, but he was made Abbot of Glastonbury, and after his master's murder, he became the most powerful man in the kingdom.

Glastonbury was then the Mecca of England. To its shrine Athelstan and Canute made pilgrimages. Within its sacred precincts were buried Edmund, son of Edward the Elder, Edred, and Edmund Ironside.

Of much more interest and importance, however, than any of these points is the association of Somerset with King Alfred, not only the most heroic figure of his time, but one of the greatest men in the history of the world. There are few historic spots in Britain better known by name than the Isle of Athelney, near the meeting-point of the rivers Tone and Parrett, where the king spent three months of waiting, maturing his plans and determining his method of attack, before his great victory over the Danes at Ethandune, in 878.

The Isle of Athelney of to-day is an insignificant rising, barely two acres in extent, hardly noticeable in the great alluvial plain in the heart of Somerset. In Alfred's time it was the one spot of dry ground in an almost impassable swamp closely covered with forests of birch and alder. Nearer to the actual meeting-place of the rivers is a steep conical hill, called by the moor folk the Burbridge Mump. It is connected by a ridge with Athelney, and is believed to have formed an important part of the king's defences.

Burrowbridge and Causeway towards Athelney

It must always be remembered that, although the country was at that moment in the gravest peril, that although, save for this one remote spot in the Somerset marshes where the torch of freedom still burnt clear, the Danes were masters of all England, King Alfred was here no solitary fugitive, in fear of his life, reduced to hiding himself in a herdsman's hut, where he let the unregarded cakes burn upon the hearth. Such legends grew up after his day. Chroniclers of his own time tell a very different story. They describe him as a dauntless leader of men, not only holding his own in his unassailable stronghold, but pressing hard upon the enemy. Where the battle of Ethandune was in which he overthrew the Danes is a matter of some doubt. Historians are accustomed to place it at Edington, near Westbury in Wiltshire. It has, however, been suggested, with considerable show of truth, that it was at another Edington, in the Polden Hills, where may still be traced some earthworks which may be the remains of Guthrum's camp.

It is believed that the font in the little church at Aller is the very one in which, after the battle, the Danish leader and thirty of his captains were baptized. And at Wedmore, a few miles away, where the foundations of King Alfred's summer palace have been laid bare, was signed a treaty whose effects were felt even on the continent of Europe, where it was recognised that the power of the Danes was broken, at any rate for the time.

There was much trouble from them in Somerset, however, despite Alfred's long-continued and victorious conflict. Again and again they and other pirates plundered this coast, landing particularly at Uphill, Porlock, Watchet, Minehead and at the mouth of the Parrett. Victory was not always with the invaders. In 918 they were beaten off from both Watchet and Porlock, and escaping with difficulty, they established themselves for a time on Steep Holm, where they were blockaded, and reduced to great extremities for want of food. Uphill and Hobb's Boat are believed to have been named after Hubba, who, having landed near Bridgwater in 878, not long before the battle of Ethandune, was killed in fight near the shore of the Parrett.

Saxon Font at Aller

There are several other vestiges of these Danes or Norsemen in place-names in Somerset. "Holm," meaning an island, is a Norse word. So is "wick," a creek or bay, in Wick St Lawrence. And in "Longford

Grounds," the name of a shoal near Clevedon, we may perhaps trace "Long Fjord," the Norse name for the Bristol Channel.

An interesting and significant point in connection with the Danish invasions of Somerset lies in the fact that in the Royal collection of coins at Stockholm there are many more specimens of some Somerset-struck Saxon coins than there are even in the British Museum. This is especially the case with the silver pennies of Ethelred II,

Silver Penny of Ethelred II
(*Taunton Castle Museum*)

Obv. ✠ ÆÐELRÆD REX ANGLO
Filleted head to r. with Sceptre within inner circle
Rev. ÆÐELRIC M-O BADA
Hand of Providence between A and ω (Alpha and Omega)

who in vain paid the Danes to leave the country; and they probably represent, partly the plunder of this coast, and partly the bribes of the unfortunate king, whose surname of Unradig, that is to say, "He who does not take counsel," or "The Headstrong," has been mis-rendered "The Unready."

In 1067, when William the Conqueror was besieging Exeter, Harold's mother Gytha, with many nobles and their wives, escaping from the beleaguered town, took refuge for a time on Steep Holm, before making their way to Flanders. And in the following year Harold's sons sailed up the Avon and plundered the north-east of Somerset, a raid whose effects have been traced in the diminished value of many manors in that district, as given in Domesday Book.

The Saxon population of Somerset did not readily submit to Norman rule. A network of fortresses, placed in the hands of some of the Conqueror's most trusted knights, was found necessary to hold the stubborn west countrymen in check, and the last fight for freedom, in Somerset, was the Saxon attack on the impregnable castle of Montacute. By the time of the Domesday Survey very few English landholders remained, an important exception being the Abbot of Glastonbury. One-fifth of the whole county having belonged to the Crown before the Conquest (Edward the Confessor is named as the owner, William did not recognise Harold), is set down as the property of the new king, in the description of each of whose manors are the words "It has never paid Danegeld, nor is it known how many hides there are in it." An interesting survival from Norman times is the ringing of the curfew bell, which is still kept up at Taunton and other places.

In the Civil War of Stephen's time, some of the Somerset nobles took the side of Queen Maud, and from their strongholds of Rougemont at East Harptree, of Dunster, and of Castle Cary in particular, they wasted the lands of the king and his adherents. So completely did De Moion of Dunster defy the Royal authority that he went the length of coining his own money, specimens of which are still in existence.

None of the well-known battles of the Wars of the Roses was fought in Somerset; but it was during that period, in 1457, that Stogursey Castle was taken and burnt, not, apparently, because of any leaning towards either York or Lancaster, but because it was the stronghold of a troublesome robber-baron. It was also during these wars that Wells Cathedral and its precincts was surrounded by a fortified wall, some of whose gates still remain.

Near the close of the fifteenth century, in 1497, the county was crossed by two rebel armies. The first, composed of Cornishmen who had risen in arms against a heavy war-tax, was ultimately defeated at Blackheath. The second, one of whose commanders was Lord Audley of Nether Stowey, was acting in support of Perkin Warbeck, who pretended to be that Richard Duke of York who was said to have been murdered in the Tower. Henry VII himself took the field against Warbeck, who fled at the approach of the Royal forces, but finally surrendered to the king at Taunton.

The suppression of the monasteries by Henry VIII was nowhere more severely felt than in this county, in which there were no fewer than 50 communities of monks and nuns. The famous abbey at Glastonbury was the oldest, largest, richest, and most powerful monastic establishment in the island; and the execution of Whiting, the last of its Abbots, on Glastonbury Tor, is a dark blot on Henry's reign.

The great event in Somerset in Queen Elizabeth's time was the preparation, begun quite early in her reign, to meet the expected Spanish invasion. The forces of this county were declared to be among the most efficient in the kingdom. Somerset provided several ships for the navy, and when the first news that reached England of the actual sailing of the Armada was brought into Bridgwater, on the 21st of July, 1588, four thousand well-armed Somerset men marched up to London to help in keeping out the Spaniards.

Somerset was the scene of some of the most important events in the Civil War. It was, indeed, in this county that the first conflict took place. On the 2nd of August, 1642, or 20 days before the king raised his standard at Nottingham, 600 Parliamentary infantry, while marching to capture a small body of Royalist horse, were ambushed by eighty Cavaliers, at Marshall's Elm, near Street, and completely routed, with a loss of twenty-five killed. The subsequent fighting included two important battles, several sieges, and many minor actions and skirmishes. At first the county as a whole was for the Parliament. But after the Royalist victory at Lansdown, in 1643, it declared for King Charles, who, in that year, himself marched through Somerset. But the tables were soon turned. In 1644 and 1645 the Parliamentary army carried all before them. The heroic defence of Taunton by Blake against the Royalists, who made most desperate efforts to retake the town, is one of the most stirring episodes of the whole war. On the other hand Bridgwater, although well fortified and abundantly supplied with provisions and ammunition, and expected to hold out for a long time, surrendered to Fairfax after a siege of less than a fortnight, followed by two days storming; and its loss was a great blow to the Royal cause. The most important and decisive battle was at Langport in 1645,

The Field of Sedgemoor
(*Weston Zoyland church in the distance*)

when Fairfax, by defeating Goring, finished what he considered to be the best of any of his campaigns. He was in command, but it was Cromwell who chased the flying Cavaliers through the blazing streets of the town. Dunster Castle, which, after a siege and blockade of 160 days, surrendered to Blake in 1646, was the last place in Somerset that held out for the king.

In the summer of 1685 Somerset was the scene of the last battle, worthy of the name, which has taken place on English ground. It was in this county that the Duke

Sunset on Sedgemoor[1]

of Monmouth, who had landed at Lyme, in Dorsetshire, and had declared himself the rightful heir to the throne of England, collected most of his little army. He was proclaimed king in the market-place of Taunton, and at the cross of Bridgwater, and he was received with

[1] The water is lying in the old Bussex Rhine, which except for this slight hollow, generally dry, has been filled up.

acclamation by the common people; but hardly any one of real influence or importance joined his ranks.

The night march by which he attempted to surprise the camp of the Royal troops at Weston Zoyland, four miles from Bridgwater, brought on an action called the Battle of Sedgemoor, which began between one and two

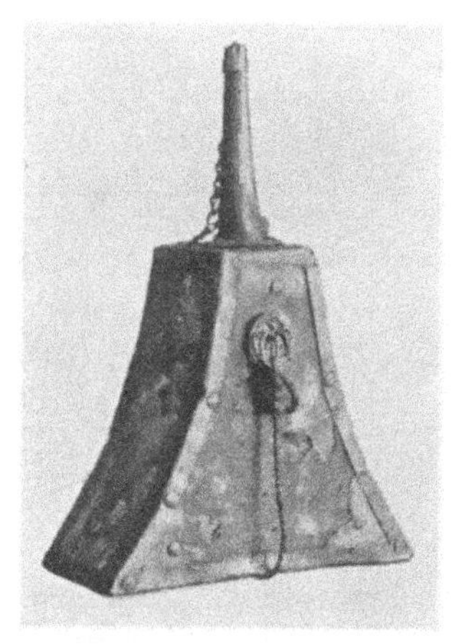

Powder-horn found at Sedgemoor
(*Taunton Castle Museum*)

o'clock in the morning of the 6th of July, 1685, and which, after an hour and a half's fighting, ended in the total defeat of Monmouth's ill-armed and untrained little army. He himself fled from the field, but was captured in Hampshire two days later and beheaded on Tower Hill.

It is stated in a manuscript account of the battle preserved in the church of Weston Zoyland, that only 16 of the king's soldiers were killed on the spot, and that some of them were buried in the neighbouring church, and some in the churchyard. The same account states that 300 of the rebels perished in the fight, and that many more died of their wounds or were killed in the pursuit. The rebel dead were buried under a great heap of sand, whose site has been determined by excavation, but it

The Grave Ground, Sedgemoor

has now so completely sunk down below the surface of the moor, that not the slightest rising in the green grass of "The Grave Ground" remains to mark the spot.

When the fighting was over, the prisoners were treated with great cruelty by Colonel Kirke; while the brutal severity of Judge Jeffreys, who, in a series of trials, conducted at various towns, sentenced more people to death than perished in the persecution of the protestants under Queen Mary, is one of the darkest episodes in English history.

Three years after the Battle of Sedgemoor, the Prince of Orange, afterwards proclaimed king as William III, marched through Somerset on his way to London; and the first blood shed in that brief campaign was spilt at Wincanton, in a skirmish between a small body of the Prince's troops and the army of King James II.

18. Antiquities—Prehistoric, Roman, Saxon.

The people who inhabited this country in prehistoric times, that is to say, before the coming of the Romans, are said to belong—taking them in the order in which they flourished—to the Stone Age, the Bronze Age, and the Iron Age. They are thus described from the material which they used for their tools and weapons. The men of the Stone Age are further divided into Palaeolithic and Neolithic—that is to say, those of the Earlier and those of Later Stone Age. The implements of the former are made of chipped and unpolished stone; while many of those of the latter and more recent races are beautifully smooth and well-finished.

These primitive races, especially those of the Neolithic Age, have left abundant relics and traces of their existence in Somerset, although it is not always easy to decide to which of the three ages a particular relic may belong. The use of flint for arrow-heads, for example, probably

Palaeolithic Flint Implement
(*From Kent's Cavern*)

Neolithic Celt of Greenstone
(*From Bridlington, Yorks.*)

continued long after the invention of bronze; and there can be little doubt that bronze was employed, especially for ornaments, long after iron weapons had come into general use. Authorities think that the Bronze Age began about 2000 years before the Christian era; and it was during the Iron Age that the Romans conquered Britain.

Bronze Bull's Head, Late Celtic (found at Ham Hill)

(*Taunton Castle Museum*)

Relics of the Stone Age, in the shape of axes, spear and arrow-heads, scrapers, borers, and even saws of flint or chert, have been found in all parts of Somerset, most of them Neolithic, and many of them showing much skill in working a very difficult material. In some cases they are associated with the bones of animals long extinct in Britain. The skeleton of a man believed to have belonged to the Stone Age, one of very few such that have ever been discovered, was found in a cave at Cheddar.

The men of the Bronze Age reached a much higher state of civilisation than their predecessors, and they left much more striking relics. There is reason to believe that it was they who set up the great stones in the circles at Stanton Drew, seven miles south of Bristol, although some authorities think them the work of a Neolithic people. These remains consist of large unhewn stones, most of which have fallen, arranged in three separate circles of different sizes. The largest circle, which probably origin-

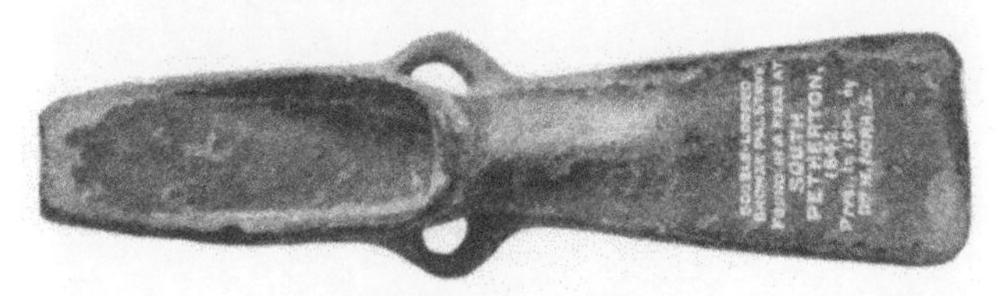

Bronze Age Weapon
(*Taunton Castle Museum*)

ally consisted of thirty upright stones, is rather more than 120 yards in diameter. Only three stones now stand erect, fifteen others are visible, six buried, the rest—if there were thirty originally—cannot be traced. The largest stones are in the north-east circle; one fallen block is 13 feet long. Two of the circles have avenues or alignments, that is to say, straight rows of stones, pointing, one of them east and the other north-east; and there is also a structure of three great flat stones called the Cove.

It is thought that such circles were probably connected with worship, and it is certain that some were used as places of burial. But one important use of them seems to have been as rough astronomical instruments for regulating the true length of the year by the observations of the rising either of the sun or of one of the stars.

The Circle at Stanton Drew

Thus, the chief circle at Stonehenge is believed by Sir Norman Lockyer to have been constructed early in the Bronze Age, or about 1600 years before Christ, for the purpose of observing the time of sunrise on Midsummer Day. The same authority thinks that the two Stanton Drew circles which have alignments were set up, one of them about 1200 and the other about 1075 years before Christ, to watch the rising of the Star Arcturus, in May; a month which, in ancient times, was regarded by some nations as the first of the year. The Cove may have served as a seat for the observer.

Another relic, which may be attributed to this or to the preceding period, is the sepulchral monument or cromlech, consisting of three great stones, which stands on a knoll in Orchardleigh Park.

The discovery of the lake-village at Godney near Glastonbury has thrown a flood of light on the little-known manners and customs of the prehistoric Iron Age. The ancient settlement, which occupied rather more than

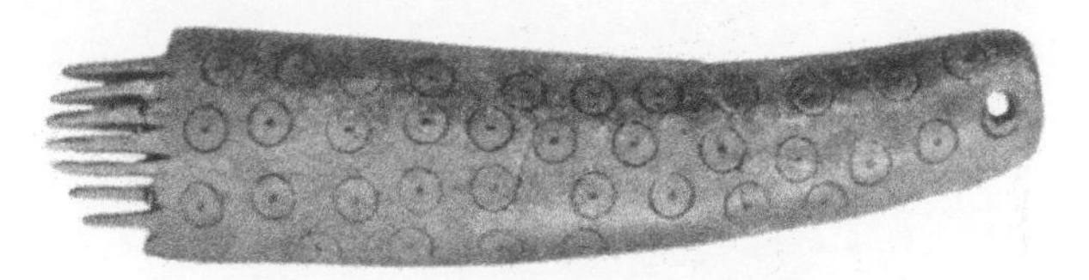

Weaving Comb from Lake Village
(*Taunton Castle Museum*)

three acres of ground on the shore of a shallow lake, long since dry, consisted of nearly 100 round thatched huts, made of wicker and clay attached to upright posts, and standing on clay platforms resting on brushwood which was kept in place by wooden piles. The whole was surrounded by a fence or palisade. A great many relics of the highest interest have been found here, showing that the inhabitants were artistic workers in bronze, iron, and even glass; that they were clever potters; that they were skilful carpenters, making looms, wheels, and canoes; that they were hunters and farmers, warriors and even horsemen; and that they possessed, if they did not make, highly-finished ornaments of jet and amber. Perhaps the most striking object is a bronze bowl, finely decorated with bold and artistic repoussé work.

Bronze Bowl from Godney Lake Village

The absence of money, except for one coin believed to belong to the first century before Christ, and the presence of great numbers of iron currency-bars, such as are mentioned by Julius Caesar, point to a time before the Roman occupation; and authorities consider that the village was inhabited from about 250 years before the Christian era, to about 50 years after it.

There are in Somerset 60 or 70 camp or hill-forts, belonging to various periods, some of them at least as old as the Bronze Age. The most remarkable are those of Worlebury, on the hill above Weston-super-Mare, and Dolbury, in the parish of Churchill, both of which were fortified with walls of loose, unmortared masonry; of

Wick Barrow, from S.E.

Cadbury, near Wincanton, believed to have been the Camelot of King Arthur; of Castle Neroche, the great fortress commanding the Vale of Taunton Dean; and of Hamdon Hill, near Yeovil, whose tremendous earthworks enclose a space of 200 acres.

On the heights of Mendip, of the Quantocks, and of Exmoor, are many tumuli or burial-mounds, a large proportion of which lie near lines of ancient road-ways. Some are in groups, such as the Priddy Nine Barrows; others, like the great mound which gives its name to the village of Rowberrow, stand alone. The most striking of them all is the tumulus of Stoney Littleton, six miles south of Bath. It is an oval mound, 107 feet long, surrounded

Beaker from the Wick barrow, Early Bronze Age
(*Taunton Castle Museum*)

by a stone wall, and containing seven burial chambers, in some of which were found bones and rude pottery. The majority of the Mendip barrows were opened early in the nineteenth century, when many curious things were found, including a bronze arrow-head, "sharp enough to mend a pen."

To the prehistoric period also belong three interesting objects on Exmoor—the bridge of great flat stones, known as Torr Steps, across the river Barle; an upright stone on Winsford Hill, on which is carved CARATACI . F, which is believed to stand for "the son of Caratacus," or Caractacus, as the name is generally mis-spelt; and

Torr Steps, on the River Barle

a stone circle, not far from the same spot. Many very fine examples of the metal-work of the Bronze Age have been found on the turf-moors and in other places, such as axe-heads and swords, and beautiful torques and other ornaments of bronze and even of gold.

Coins that were struck in Britain before the time of

Julius Caesar have no inscriptions. A few such, of gold, silver and bronze, some of them stamped with the figure of a horse, have been found in Somerset. The most remarkable piece of ancient money yet discovered in the county is one of those barbaric imitations of the Macedonian gold stater which are believed to have been made in Gaul, in the third century before Christ, for circulation in this country. This specimen, which was found at Churchill, is of gold, about the size of a sovereign, bearing on one side a horse, roughly copied from the original fine

Gold Stater of Philip II of Macedon

British Gold Coin, found at Churchill

(*Taunton Castle Museum*)

Macedonian stater of Philip II, and on the other side no device whatever.

The Roman settlements in Somerset were less purely military than those in many parts of Britain. The conquerors, as was observed in the previous chapter, were early attracted by the hot springs of Bath, which became one of their most important cities, by the mineral wealth of Mendip, and by the pleasant and fertile country in the north and east, where many traces of their occupation in the shape of camps and villas still remain. Few towns

The Roman Baths at Bath, before restoration

in England have yielded more Roman relics than Bath, which is especially rich in architectural antiquities. The Roman baths there are the largest and best preserved in

Stone Head from Temple of Sul Minerva, Bath

Britain ; while, to take one notable example, the head of a deity, boldly carved in stone, and suggestive of the head of the sun-god on the coins of Rhodes, is a most striking piece of Roman sculpture.

Several Roman roads passed through Somerset. The most remarkable of them, the Fosse Way, described in another chapter, crossed the heart of the county on its way from Lincolnshire to Devon. Another road, that ran from Old Sarum in Wiltshire to the port of Uphill,

Roman Pavement, East Coker
(*Taunton Castle Museum*)

traversed the Mendip Hills. Many Roman camps, distinguished from those of the Britons or Saxon by their regular rectangular shape, are to be seen, especially near Bath. The remains of more than 50 Roman villas, a large proportion of them along the Fosse Way and near

Bath, have been found; some of them, such as those near Yatton and at Whatley, containing very beautiful tesselated pavements.

At the ancient mining-station of Charterhouse-on-Mendip a great many Roman antiquities, such as Samian and other pottery, inscribed stones, pigs of lead marked with emperors' names, engraved gems, weapons, tools, and coins, have from time to time been dug up. Roman coins have, indeed, been found in all parts of Somerset, sometimes in hoards of many thousands, and remarkable in being of specially late date. Once, when a jar containing a couple of gallons of Roman coins was found at Yatton, the third brasses passed for some weeks, as farthings, in the shops of the village. Roman pottery-kilns have been discovered at Shepton Mallet, and moulds for making coins at Burtle, on the turf-moor.

Remains that can be definitely attributed to the Saxons are comparatively few in Somerset. It is quite probable that the troops of Ceawlin, Kenwalch, and other Saxon leaders constructed some of the many hill-forts. And the earthen rampart called the Wansdyke, which, starting at Portishead, runs through Gloucester and Wiltshire into Oxfordshire, and perhaps appears on the further side of the Thames, is believed to be of Saxon handiwork. The Saxons coined money at eleven different places in the county, and many thousands of Somerset-struck silver pennies are known.

But by far the most remarkable Saxon relic ever discovered in Somerset, one of the most interesting objects, indeed, that has been found anywhere in England, is the

"Alfred Jewel," which in 1693 was dug up four miles from Athelney, half-way between that place and Bridgwater, and is now in the Ashmolean Museum at Oxford. It consists of an oval crystal set in gold, with

The Alfred Jewel (front view)

a figure in coloured enamel showing through the stone. Around the gold setting are the words AELFRED · MEC · HEHT · GEWYRCAN ·, that is to say, "Alfred had me made." What this object was used for is uncertain. But the most reasonable suggestion is that it was the head of a pointer or reading-staff, for the master of a choir. Alfred himself, in his translation of Gregory's *Cura Pastoralis*, states his intention of placing in the see of every bishop in the realm, with a copy of the book, "a reading-staff of (the value of) fifty mancuses." The legend round the gem seems clear evidence that we have here the head of one of these very staves; that in this has survived one of the most cherished treasures of the monastery of Athelney, and a very real link with the great sovereign himself.

The Alfred Jewel (side view)

The most interesting mediaeval relic found in Somerset is the Kewstoke Cup, a decayed and broken wooden vessel which, in 1849, was found concealed in a hollow capital of Caen stone, built into the wall of Kewstoke church. There can be little doubt that this was one of the reliquaries which, after the murder of Thomas à Becket, were sold to pilgrims at his shrine at Canterbury; and that the thin, dark layer of solid matter at the bottom of it is what passed, more than seven hundred years ago, for the blood of the archbishop.

19. Architecture—(*a*) Ecclesiastical.

We will consider the architecture of the buildings in Somerset under three divisions, viz.: (*a*) Ecclesiastical, or buildings relating to the Church; (*b*) Military, or Castles; (*c*) Domestic, or houses and cottages.

There is one fact worth noting with regard to all these classes of buildings, and that is that—as indeed is universally the rule—the architecture of the county has been affected by the materials accessible. Thus we find that stone, wood, and bricks are used either because they could be easily obtained, or because of the wealth or otherwise of the builders.

Now with regard to the ecclesiastical buildings, let us consider first the churches and cathedrals, and then glance at the remains of the abbeys, monasteries, and other religious houses. The churches of Somerset are of various styles and of different ages, so that it will be well to classify them as Norman, Early English, Decorated, and Perpendicular.

Towards the end of the twelfth century the round arches and heavy columns of Norman work began gradually to give place to the pointed arch and lighter style of the first period of Gothic architecture which we know as Early English, conspicuous for its long narrow windows, and leading in its turn by a transitional period into the highest development of Gothic—the Decorated period.

This, in England, prevailed throughout the greater part of the fourteenth century, and was particularly characterised by its window tracery. The Perpendicular, which, as its name implies, is remarkable for the perpendicular arrangement of the tracery, and also for the flattened arches and the square arrangement of the mouldings over them, was the last of the Gothic styles and is peculiar to England. It developed almost simultaneously and uniformly towards the end of the fourteenth century and was in use till about the middle of the sixteenth century.

Examples of all these styles are to be found in Somerset, and the cathedral of Wells contains masterly work in all of them except pure Norman. But the parish churches of the county belong as a whole to the Perpendicular period; that is to say, they were built or rebuilt between 1377 and 1547, although in many of them there are features of earlier date, especially in their chancels and their crypts, where the latter exist.

The glory of Somerset churches is in their towers, which in many cases are of noble and magnificent workmanship, and contain details rarely found elsewhere except in some of the adjoining counties. The typical Somerset Perpendicular tower is square and western, and has, with other characteristics, a great western window with a small doorway beneath it, an outside stair-turret, striking parapets and pinnacles, and two square buttresses at each corner, ending in pinnacles.

In several churches in Somerset there is some striking Norman work, as, for example, the chancel at Compton Martin, the doorway and other details at Lullington, the chancel-arch at Stoke-under-Hamdon or East Stoke, and the arches of the so-called St Joseph's Chapel at

Norman Arch, Stoke-under-Hamdon

Glastonbury. Many other churches, such as Christon, Chewton Mendip, and Kewstoke, have fine late Norman doorways.

Very fine Early English architecture is to be seen in some of the best features of Wells cathedral, and in the churches of Barrington, Montacute, Croscombe, Tintinhull, and Congresbury.

The best Decorated work is also in Wells cathedral, and there are other fine examples in the churches of Ditcheat, Priston, Meare, and Yatton.

There is very fine Perpendicular architecture in Wells cathedral, but some of the best of the many magnificent towers of that period are those of Wrington, St Cuthbert's in Wells, St John's at Glastonbury, Huish Episcopi, Winscombe, Cheddar, Banwell, Chewton Mendip, North Petherton, St Mary's in Taunton, Kingsbury Episcopi, and Bishop's Lydeard.

The interiors of the Somerset churches often contain beautiful woodwork, in the shape of roofs, rood-screens, pulpits, and bench-ends. The open wooden roofs of Martock, Shepton Mallet, Somerton, and St Mary's, Taunton, are especially famous. The rood-screens scarcely equal the magnificent examples for which Devonshire is so famous, but there are very fine ones at Dunster, Minehead, Croscombe, Banwell and Norton Fitzwarren. Good bench-ends exist at Bishop's Lydeard, Croscombe, Curry Rivel, Kingsbury, and Trull. Those at Clapton-in-Gordano are the earliest in England, but on the other hand the very fine ones at Rodney Stoke are quite modern. There are fine wooden pulpits in many places; they are, indeed, a special characteristic of Somerset churches. Good examples may be seen at Bridgwater, Long Sutton, Queen Camel, North Petherton, and Croscombe. In the

Rood Screen, Banwell Church

last-named church the woodwork in general is regarded as among the finest in existence. Beautiful stone pulpits are to be seen at Cheddar, Banwell, Locking, Wick St Lawrence, and elsewhere.

There is not much stained glass of striking excellence in Somersetshire, but there are some very beautiful windows in Wells cathedral, as well as at Cheddar and Kingsbury Episcopi. Sepulchral brasses are comparatively few in number, perhaps because the excellent and easily worked freestone and alabaster of the county lend themselves so well to decoration. There are, however, some very fine brasses at Ilminster (where are the effigies of the Wadhams, founders of Wadham College, Oxford), at South Petherton, Beckyngton, St Decuman's, Churchill, and Fivehead. At the last-named is a very remarkable palimpsest brass, which has been mounted so as to show both sides. There are many recumbent stone effigies of clerics in Wells cathedral; and other remarkable monuments of a similar character, some of them of great historic interest, are to be seen at Hinton St George, Cothelstone, Rodney Stoke—where the Rodney Chapel seems crowded with monuments—Dunster, Brympton D'Evercy, Long Ashton, and Yatton. Many churches of Somerset are famous for their peals of musical and mellow bells, a few of which were cast in the century before the Reformation. Stone crosses are very numerous, and are a characteristic of the county. There are specially good preaching or churchyard-crosses at Bishop's Lydeard, Crowcombe, Chewton Mendip, and Doulting, while fine covered and market-crosses exist at Cheddar, Shepton Mallet, Dunster, Ilchester, and Martock.

By far the most striking edifice in Somerset is Wells cathedral, which, according to Freeman, "is the best

Churchyard Cross, Stringston

example to be found in the whole world of a secular church with its subordinate buildings." It is one of the smallest, but at the same time one of the most beautiful of English cathedrals, and it is, moreover, the only one

Wells Cathedral, the West Front

which is complete in all its parts. Founded originally in 704 by King Ina, it was partly re-built by Robert, first Bishop of Bath and Wells, and was consecrated by him in 1148. No definite trace of this building, however, remains, except the font. The magnificent structure that we see to-day was designed and begun in the last quarter of the twelfth century, by Reginald de Bohun, (1174—1191), whose work—in part of the nave and especially in the north porch—shows traces of the period of transition from Norman to Early English. It is thought that he also built the western end of the choir.

The main builder of the cathedral, however, was de Bohun's successor, Reginald Fitz-Jocelyn (1206—1242). He finished the nave, and erected the marvellous west front, which, with its figures of angels and apostles, of kings and prelates and heroes, and its biblical scenes, is one of the architectural wonders, not of Wells only, but of Europe. The chapter-house, the Lady-chapel, and the three eastern bays of the choir are brilliant examples of the best work of the Decorated period. The upper parts of the north-west and the south-west towers, the clerestory, the tracery of the nave windows, and the west and south cloisters are Perpendicular work. The inverted arches under the tower, which were built about 1338, when the tower seemed in danger of falling, form a very peculiar feature of the interior.

Freeman the historian considered that there was not in all Europe another such group of buildings as that on which the eye looks down from the Tor Hill on the Shepton Mallet road: "the Cathedral as the great centre, the Palace, the Cloister, the Chapter-House, the Vicars' Close, the detached houses of the canons, the more distant

Windows of Decorated Period, Wells

view of the Parish Church." Nowhere else have such buildings suffered so little from the hands of fanatics or restorers. The cathedral is the cathedral of seven hundred years since. The Bishop lives in the palace that was finished before the last Crusaders were driven from Jerusalem. The Dean inhabits the house that was rebuilt by Gunthorpe more than four centuries ago.

Wells Cathedral—the Nave, looking East.
The Inverted Arches

The dwellings in the Vicars' Close, though more altered than the other buildings, have been inhabited since the battle of Crecy.

Scattered up and down over Somerset are the ruins of many monasteries, priories, and nunneries which were closed by order of Henry VIII. These monastic houses were originally founded as places to which people might retreat who wish to retire from the world, and to lead lives of holiness, benevolence, and poverty, serving God and benefiting their fellows. For a time the inmates did all these things. As long as they were poor they were a blessing to the countries where they lived. They preached to the people, they taught in schools, tended the poor and the sick, practised agriculture and many useful arts, such as the construction of clocks, keeping alive such learning as there was, and making beautiful manuscript copies of the Bible and of the works of classical authors which otherwise would have been lost. But when they grew rich they became idle, careless, ignorant, and wicked, and their lives too often a scandal to the world. Henry VIII, as the result of a commission which he sent round to enquire into their condition, resolved on suppression. The houses were closed, their inmates scattered, their estates sold for trifling sums or given to the king's favourites, while part of their wealth was used in founding grammar-schools.

By far the most important of all these monastic houses was the Benedictine Abbey of Glastonbury, whose buildings covered 60 acres of ground, and whose mitred Head could, so it is declared, call 15,000 fighting-men to his standard. According to the story long implicitly received by the Roman Church, and still, by some authorities, regarded as authentic, it was originally founded by Joseph of Arimathea and his companions, who, having fled from

Palestine to escape persecution, landed here, bringing with them the Holy Grail, and built a tiny church of osiers, the humble forerunner of the most powerful, the richest, and the most learned of all the monasteries in the island. Here were buried some of the greatest of the Saxon kings. Here, according to the old tradition, were laid the bodies

Interior of "St Joseph's" Chapel, Glastonbury

of Arthur and Guinevere. The most famous of its long line of abbots was Dunstan, the foremost Englishman of his time. The last was Whiting, who, after a mock trial at Wells, on a trumped-up charge of stealing the abbey plate, was hanged on the Tor, the hill that looks down on the wreck of the great house that he had governed.

Of the great Abbey there remains to-day only a beautiful fragment of ruin, whose chief features are the so-called St Joseph's Chapel—which was really the Chapel of St Mary—the walls of the choir, the finely-moulded

Glastonbury

arches of the Lady Chapel, and the piers of the great central tower. Not far off is the large and curious abbot's kitchen. All else of consequence has been destroyed. For ages the walls of the abbey served as a quarry for the township, half of whose houses are said to have been built of its costly stones, which were even used to make the road between Glastonbury and Wells.

Gateway, Cleeve Abbey

The fate of Glastonbury was the fate of fifty other Somerset monastic houses. Of some there remain the ruins. Others have fared still worse, and have entirely disappeared. There are picturesque and beautiful old ruins at Cleeve, which belonged to the Cistercians; at Muchelney (Benedictines), at Woodspring (Augustinians), at Hinton Charterhouse (Carthusians), at Barlynch (Cistercians), at Montacute (Cluniacs), and at Stavordale (Augustinians), several of which have happily escaped destruction through having been used as dwelling-houses.

Chapter House, Cleeve Abbey

On the other hand there are several, like the much more ancient monastery which King Alfred established at Athelney, of which not one stone stands upon another; while of others, such as the Carthusian cell at Charterhouse-on-Mendip, the very site has been forgotten.

20. Architecture—(*b*) Military.

As has already be pointed out, there were in Somerset many primitive castles or fortresses, consisting simply of enclosures surrounded by ramparts of earth or loose stones, to which, in some cases, was added a wooden palisade. Such, no doubt, was the original castle built at Taunton by King Ina. After the Norman Conquest strong stone castles were erected in all parts of England, partly by order of the king himself, and partly by his barons, who found it necessary to defend themselves against the Saxons whose lands they had seized. By the end of the reign of King Stephen there were 1115 such strongholds in England.

A Norman castle, which sometimes occupied a space of many acres, was usually built on high ground or close to a river, and was surrounded by a ditch or moat which, if possible, was filled with water. Inside the moat was a high and very thick wall, generally with towers at intervals, with a parapet to shelter the men fighting on the top of it, and with spaces, called embrasures, through which they could shoot arrows at the enemy. The principal gate was very strongly defended by covering-towers, and above it were holes through which melted lead, or boiling water or pitch, or hot sand could be poured on the besiegers. It was reached by a drawbridge, and was closed by a strong door and a portcullis. Smaller gates were the postern and the sally-port. The space inside this wall was called the outer bailey. Inside it was another wall, also with towers and a gate, enclosing the inner bailey, in which were dwellings and store-houses. Within the inner bailey was the most important part of the fortress, a high tower called the keep, a building of many floors, with walls 15 or even 20 feet thick, the last place of retreat when the rest of the castle was taken. On the ground floor, which had no windows, were the well, the dungeon and the store-rooms. On the next floor, which was lighted only by narrow loop-holes, were the soldiers' quarters. On the upper floors were the chapel and the rooms of the governor and his family.

There were at least twelve such castles in Somerset: Montacute, Castle Cary, Dunster, Stoke Courci or Stogursey, Somerton, Bridgwater, Nunney, Farleigh Hungerford, Richmont, Taunton, Nether Stowey and Castle Neroche, of which the first four, two of them inland and two by the sea, formed a sort of Quadrilateral, with which the others were more or less connected strategically. All twelve have played their part in history; some in the Conqueror's time, some in the stormy days of King Stephen, some in the struggle between the King and the Commons in the great Civil War. And of the whole twelve Dunster is the only one which has not been dismantled or destroyed. Of the great stronghold of Montacute, built during the reign of William I on a hill near the village, and once a place of especial strength and importance, nothing now remains.

All that is left of Somerton Castle is a piece of walling that has been used in the building of an inn. Of Castle Neroche not one stone stands upon another.

Of the castle of Nether Stowey, once the home of the Lord Audley who led the Cornish insurgents in 1497, and who was afterwards beheaded on Tower Hill, nothing now exists but some grass-covered mounds on a hill near the village. Excavations made in 1890 in the mound at Castle Cary laid bare the foundations of a very large keep, 78 feet square, and with walls 15 feet thick; the largest but four in the whole country. But the fortress is not mentioned in history after the civil wars of Stephen's time, when it was very likely destroyed. Another castle figuring in that struggle was Richmont, or Rougemont, at East Harptree, held by Sir William de Harptree for Queen Matilda, and taken by storm by Stephen himself. It was pulled down by its owner in the reign of Henry VIII, and its materials were used to build a more convenient dwelling. Little more of it remains than the foundation of the keep, and some part of the fortified way down to the stream below.

Stogursey, originally Stoke Courci Castle, built on low ground and surrounded by a moat, was the stronghold of the family of de Courci, of whom one conquered Ulster, while another descendant, Baron Kingsale, is the holder of the oldest peerage in the United Kingdom. The castle passed in the thirteenth century to Fulke de Breauté, on his marriage with an heiress of the de Courcis. He became such a terror to the neighbourhood that his fortress was dismantled by order of Hubert de Burgh,

Chancellor to King Henry III, but it was restored in the following reign. During the Wars of the Roses it was taken from one of the Percys by William de Bonville, and burnt, and it has ever since remained a ruin. It was a small building, and its scanty remains consist of little more than the bases of the covering-towers, part of the sally-port and the lower courses of the keep.

Stogursey Castle

Bridgwater Castle was built in 1216 by William de Briwere, one of the most remarkable men of his time. It was a large and important fortress, but it has now almost entirely disappeared. It was in good repair at the time of the Civil War, when it was held for King Charles. But although strong, well-garrisoned, and amply furnished with stores, so that it was expected to make a long defence, it was taken by Fairfax, in the summer of 1645, after a desperate storm of two days. It was so completely dismantled by order of Parliament that there is now nothing left of it but some vaults and a fine stone gateway on the left bank of the river Parrett, believed to have been the water-gate of the castle.

Nunney Castle, which appears to have been fortified by Royal licence in the fourteenth century, was taken by Fairfax in the same campaign. But although dismantled, it was not destroyed; and its ruin still stands, a severely plain oblong building, with a tower at each corner. It suffered little damage in its one day's siege, and time has lent it no little picturesqueness.

Taunton Castle, originally founded by King Ina in 702, and no doubt then consisting of earthworks and timber, was built in stone by the Bishop of Winchester in the time of Henry I, and was one of the largest fortresses in Somerset, occupying several acres of ground. It was taken by the Cornish insurgents in 1497, when they dragged out of it and hanged the King's Commissary, whom they associated with the heavy taxes of which they complained. Here, too, in the same year, Perkin Warbeck, who had been brought from sanctuary at Beaulieu, surrendered to Henry VII. By far the most stirring episode in its history is Blake's heroic defence against the Royalist army in 1645. There are considerable remains of the castle, and it is now occupied by the Somersetshire Archaeological Society as their Museum and Library. In its great hall Judge Jeffreys held his infamous assize, after the battle of Sedgemoor.

Farleigh Hungerford Castle is a picturesque and interesting ruin, very strongly situated above a steep slope by the river Frome. Originally a dwelling-house, its conversion into a castle was begun in the time of Richard II by Sir Thomas de Hungerford, and finished by his son Sir Walter, who fought at Agincourt, and having, it is said, taken prisoner the Duke of Orleans, spent that nobleman's ransom in completing his fortress. A later Sir Walter, of Henry VIII's time, kept his third wife a close prisoner here for four years, making several attempts to poison her. He was, however, beheaded. His widow married again, and survived him 31 years. In the chapel are many monuments, among them the recumbent effigy of the founder of the castle, who died in 1395. Farleigh was held for King Charles, but surrendered to the Parliamentary army in 1645. It was not dismantled, but has not been inhabited since 1730.

Dunster, the only Somerset castle which remains entire, was originally built by William de Moion, one of the most distinguished of the Conqueror's knights. His descendants held it until 1376, when the widow of the last of the line sold it to Lady Luttrell, niece of the Black Prince; and her descendants have owned it to the present day. King Stephen tried in vain to take it, and in 1646 it only surrendered to Blake after a siege and blockade of 160 days, when its garrison marched out with all the honours of war. It was the last place in Somerset to hold out for King Charles. Parliament directed that it should be dismantled, but the order was happily disregarded.

Dunster Castle stands in a commanding position, among most beautiful and picturesque surroundings. The oldest parts of it are the gateway and its covering-towers, which were built in the thirteenth century. But the general structure has been greatly altered at various

High Street, Dunster, showing Castle and Market Cross

periods, especially in its interior, and it now ranks as one of the finest and best-appointed as well as one of the most picturesquely-situated private residences in Somerset.

Old Military Stables, Dunster Castle

21. Architecture—(*c*) Domestic.

The great houses of Somerset are among the finest in England, and although many of them have lost their ancient dignity and are occupied now only as farm-buildings or have fallen into ruin, there still remain, scattered up and down over the county, many noble examples of Plantagenet, Tudor, and Jacobean domestic architecture. There are so many of the two latter, at all events, that there is hardly a parish in all Somerset in which there is not a house at least as old as the days of Queen Elizabeth; and it will be impossible here to do more than allude to a few typical examples.

The finest thirteenth-century residence in our island is the Bishop's Palace at Wells, which, although altered and added to by many hands in the ages that have intervened, remains much as it was originally built by Bishop Jocelyn, between 1206 and 1242, and exhibits many striking and beautiful features. The great room in the basement has a fine ribbed and vaulted roof of stone, supported on pillars. The noble west gallery above it, and the living rooms on the same floor—an arrangement common in mediaeval houses—are lighted by the original Early English windows. The chapel is of later date, and, like the once magnificent great hall adjoining it—dismantled in the sixteenth and seventeenth centuries, and now a ruin —is a fine example of Decorated work. The palace stands in a fortified enclosure, with embattled walls, gate-house, drawbridge, and moat, dating from the first half of the fourteenth century.

Ruins of the Great Hall, Bishop's Palace, Wells

The most remarkable fourteenth-century house in Somerset is Clevedon Court, which has been called "the most valuable relic of early domestic architecture in England." It is further interesting as the "Castlewood" of Thackeray's *Esmond*. Although much altered in the reign of Queen Elizabeth, it still retains many fine features of the time of its original erection in the days of Edward II. Other good fourteenth-century residences are the manor house at Martock, an extremely fine example, with an open timber roof; the very picturesque house called Lyte's Cary, now a farm, and with its great hall in use as a cider-cellar; the abbot's house at Meare, also a farm, and with its noble banqueting-hall converted into a cheese-room; and the ruinous but beautiful manor house at Clapton-in-Gordano.

Buildings of the fifteenth century are more numerous, although many of these, again, have been converted into farm-houses, while not a few have fallen into decay. Fine examples of this period are the Deanery at Wells; the George Inn at Glastonbury, whose front, dating from the time of the Wars of the Roses, is a magnificent piece of panelling; the Abbot's kitchen in the same town,—a perfect little building, with great fireplaces in the four corners running up into a curious pyramidal central chimney; the Abbot's house at Muchelney, a beautiful building, nearly perfect, and presenting a specially fine example of a nobleman's house of the period; the manor house of Hutton; of Croscombe, partly used as a Baptist Chapel; of Chew Magna, now a school-house; and lastly Tickenham Court.

The George Inn, Glastonbury

A large proportion of the Somerset houses were built or almost entirely remodelled in the days of Henry VIII and of Elizabeth. By far the finest of these is the magnificent mansion of Montacute, built from his own designs, between 1580 and 1601, by Sir E. Phelips, sometime Speaker of the House of Commons, whose descendants have held it ever since. Grand as the house is, a masterpiece of the architecture of its time, it owes much to its fine situation and to its beautiful gardens. Its east and west fronts are both extremely fine, and the former contains no fewer than 41 Perpendicular windows. It has a noble hall, with a musicians' gallery, and the upper floor is chiefly occupied by the picture-gallery, a magnificent room 184 feet long, its oriel windows making it longer than the main part of the house. Over the eastern doorway of the mansion is the famous motto :—

Through this wide-opening gate,
None come too early, none return too late.

Over the western doorway is the almost equally familiar :—

And yours, my friends.

In 1645 Montacute was sacked by the Parliamentary Army, and was occupied for a time by Cromwell himself.

Another fine Tudor mansion is Barrington Court, which was built earlier in the period. Although greatly damaged by neglect, shorn of much of its interior decorations, and with hardly any ornament beyond its very curious twisted chimneys, it is a striking, though plain piece of architecture.

The Abbot's Kitchen, Glastonbury

Other fine Tudor houses are at Brympton D'Evercy, a noble residence, whose beautiful little private chapel contains some very interesting monuments; Hinton St George, once the residence of the Sir Amyas Poulett, who put young Wolsey in the stocks, and whose grandson was one of the guards of Mary Queen of Scots; Combe Sydenham, where Sir Francis Drake's wife lived and where he himself is said to have resided for a time; Portishead Manor; Dodington Court; Lower Halsway, the hunting-lodge of Cardinal Beaufort; Nailsea Court; and East Quantockshead, a manor held by a direct descendant of the Radulphus Paganall who is named as its owner in Domesday Book. It is interesting to remember that "Bess of Hardwick," that famous builder of houses, had a hand in one Somerset mansion, Sutton Court, a noble house, dating in part from the time of Edward II, to which, in 1558, she added a chapel and a "grand parlour."

Some good seventeenth-century houses are Cothelstone Manor, Chelvey Court, now in ruins, Tilly Manor and Gournay Manor, both now occupied as farm-houses. One owner of the latter, Sir Thomas Gournay, stood by when Edward II was murdered, and having fled the country, was taken, and beheaded at sea.

The abundance and variety of good building-stone is one of the reasons why Somerset possesses so many of these fine old houses. As might be expected, the character of the domestic architecture of the county varies as a rule, with the material most ready to hand. The beautiful brown oolite of Ham Hill, where there are extensive

Montacute House

quarries, has been much used in the south, both for houses and churches. Bath city is largely built of the paler freestone of its own district. The houses in Bridgwater again, where stone is scarce and clay abounds, are chiefly of brick,—a fine material, to which time often lends great beauty. On the moors, many of the cottages are built of mud. Many modern houses in the Mendip country, and many west-country railway stations, are built of a warm-coloured conglomerate, much quarried at Draycott, near Wells.

Perhaps the typical Somerset house of moderate size is of stone—lias or sandstone or carboniferous limestone, according to the district—covered with rough-cast and whitewashed. There are some fine half-timber houses in the west and south, especially at Yeovil, and there are also some good examples at Wells. Thatch, once the common roofing-material, has largely given place to tiles, which are made in great quantities in the county. The result has been an undoubted loss in picturesqueness, but an equally certain gain in comfort and in freedom of danger from fire. There are still, however, many thatched houses and cottages in the central districts of Somerset.

22. Communications—Past and Present.

In prehistoric times Somerset was crossed by a network of trackways, some of which are to-day broad and well-kept roads. Others, that once served to connect one hill-fort with another, and are now hardly to be traced except by the burial-mounds which, after the manner of the ancient inhabitants, were so often raised by the road-side, have long since fallen out of use. These roads, probably began in the Neolithic Age as footpaths, were made into pack-horse roads in the Bronze Age, and were more or less adapted for wheeled traffic by the prehistoric users of iron. Many of them ran along high ground, following the lines of hills, and linking together the earthworks that were not so much the dwelling-places of the ancient tribes as their camps of refuge in times of danger.

A good example of an ancient road is that which, coming out of Wiltshire, traverses the Mendips, by way of Charterhouse, Shipham, and Banwell, on its way to Uphill and the sea. Although certainly used at a later period by the Romans, it is clearly marked as of prehistoric origin by the forts and burial-mounds near its line. Another such road, whose origin is proved in a similar way, follows the line of the Quantock Hills. Both these roads are now in great part disused, but the ancient roads along the top of the Poldens from Dunster to Lynmouth, and from Bath to Wells over the Mendips, are examples of prehistoric lines of traffic which are still in use.

In common with other counties Somerset possesses many hamlets whose names end in "ford," showing where the old roads passed through the shallows of streams or rivers, such as Allerford, Otterford, Mudford; while the surname Ford, taken from one of many such spots, is of common occurrence in the county.

The remarkable prehistoric bridge on Exmoor, called

Torr Steps, has been alluded to in the chapter on Antiquities. Not many of the old narrow pack-horse bridges remain, but there is a very picturesque one—belonging however to a much later period—near Holnicote, which happily has been spared to us.

The Romans, those great road-makers, left plain marks on the means of communication in Somerset, although their undoubted roads are few. The Fosse Way, said to be the most characteristic of all their great military roads, ran through the county, crossing the Mendips at Shepton Mallet, and linking together, on its way, the Roman settlements of Bath and Ilchester. It consisted of five layers or courses, which were, in all, three feet thick. At the bottom was a layer of earth eleven feet broad. Above this came five inches of sand and stones, then fifteen inches of concrete, then ten and a half inches of pounded stones and lime, and lastly, paving-stones cemented together with lime. On these may still be traced the tracks of wheels, two feet nine inches apart. The finished road was seven feet wide, and it had a ditch on each side of it. Roman roads were narrow and had passing-places at intervals.

Other Roman roads led from Bath to the mouth of the Avon, and from Bath towards London. The British road along the top of the Mendips was adopted by the Romans, who defended it with camps and guard-stations. Another Roman road branched off from the Fosse Way at Ilchester, and led to Dorchester. Still another has been traced from Street to Ilchester, and may have joined the Mendip lead mines. Other roads which have been attributed to the Romans are not accepted as such by experts. One sign of the great antiquity of a road is when it serves, as is often the case in Somerset, as part of a parish boundary.

Some of the villages in the neighbourhood of Glastonbury, built on slight risings in the moor, were formerly connected by what would be known in America as "corduroy" roads, that is to say, made of logs, of birch or alder, laid side by side upon the soft and yielding ground. The best known example is that called the Abbot's Way—though its age and its makers are quite unknown—which connected Glastonbury with Meare and Burtle, but which is now, in some places, seven feet below the surface of the moor. Some of the logs of which it is made are still sound and hard, although the road has probably been abandoned for three or four centuries.

Somerset is probably as well provided with roads as most counties, and since local government began they have been kept in good repair. There is one particular highway, that from London to Plymouth, which, after traversing the heart of Mendip, runs through Bridgwater, Taunton, and Wellington, along which passes all the "record-making" pedestrian, cycle, and motor traffic between the Land's End and John O'Groat's House. This was formerly a great coaching road, and was much improved in 1829, when the stage coach was in its glory.

Somerset canals have not proved successful as means of communication; partly because there are no great manufacturing centres to provide traffic, and partly because what traffic there is has been absorbed by the railways. Three canals, the Somerset Coal Canal, the Glastonbury and Highbridge Canal, and the unfortunate Bridgwater and Chard Canal, which was only just finished when the railway was made, and was practically hardly used at all, have been abandoned. Two canals only, the Bridgwater and Taunton Canal, and the Kennet and Avon Canal which runs for a short distance through the north-east corner of the county, are in working order, but both are carried on at a loss.

There is a little traffic on some of the rivers, and even some of the "rhines" are used by flat-bottomed boats, in which hay and reeds are brought in from outlying fields.

Two great lines of railway cross Somerset, the Great Western and the London and South Western, of which the former is by far the more important to the county. This railway, originally the Bristol and Exeter, opened as far as Bridgwater in 1841, completed to Exeter in 1844, and amalgamated with the Great Western Railway in 1876, was made by Brunel as a broad-gauge line, though its gauge has now been altered to that of most other railways in the kingdom. It is characterised by long stretches of straight and level line; and was designed, as far as possible, to avoid the hills. Starting at Paddington, the Great Western Railway runs to Bristol, and passes through Weston-super-Mare, Bridgwater, Taunton, and Wellington, on its way to Plymouth and Penzance. Along it run the famous express trains known as the "Flying Dutchman" and the "Cornishman," the latter of which, now avoiding Bristol altogether, makes no stop between London and Plymouth, a distance of 225 miles, performing the journey in seven minutes over four hours, which means an average speed of 55 miles an hour.

The London and South Western Railway, on its way from London to Exeter and Barnstaple, runs for a short distance through the south of Somerset, entering the county at Templecombe, and leaving it near Chard. On account of the route being more direct, a traveller from London may reach Plymouth in shorter time than is possible by the Great Western. The Somerset and Dorset Railway, originally called the Central Somerset, is worked jointly by the Midland and the London and South Western Companies. On its way from Burnham to Templecombe it crosses the district known as Turf Moor, where the ground is so soft and yielding that it is said that the embankment, which had to be built on piles, sank several feet after its completion.

23. Administration and Divisions—Ancient and Modern.

In the days of our ancestors the Anglo-Saxons, Somerset was governed much in the same way as it is governed now. That is to say, while the people had to obey the laws that were drawn up under the direction of the king, they had a great deal of what we now call self-government. Every little group of houses in Somerset had its own "tun-moot" or village council, which made its own by-laws (from the Danish *by*, a town) and managed its own affairs. The large divisions of the county called Hundreds—groups of a hundred families—had their more important "hundred-moot"; while the general business of the whole shire was conducted by the "shire-moot," with its two chief officers, the "ealdorman," or earl, for military commander, and the "shire-reeve" for judicial president. These three assemblies may fairly be said to correspond to the parish councils, the district councils, and the county council of the present day. Our lord-lieutenant corresponds to the ealdorman of other days, and the present sheriff to the ancient shire-reeve.

The division called a Hundred may have been so named because it contained a hundred families. But the present Somerset Hundreds, of which there are 40, are not quite the same as those of Anglo-Saxon times, and vary very much in size and population. The Hundred of Winterstoke, for example, contains 31 parishes, while the single parish of Martock constitutes a Hundred by itself.

The parish is another ancient institution, and was originally a "township or cluster of houses, to which a single priest ministered, to whom its tithes and ecclesiastical dues were paid." Many of the 480 Somerset parishes roughly correspond to the manors in Domesday Book; but the whole country was not divided up into parishes until the reign of Edward III. Queen Elizabeth made them areas of taxation, partly, at any rate, to provide funds for the relief of the poor. In modern times, with the idea of taking better care of the poor, the parishes have been grouped together in Poor-Law Unions, of which there are seven in Somerset, each provided with a workhouse, which was meant to be a place in which the able-bodied poor might find work. Now, however, they are little more than refuges for the destitute, the idle, and the incapable.

The local government of Saxon times was swept away by the feudal system of the Normans, which transferred the power of making and carrying out laws from the freemen to the lords of the various manors, and was only restored as recently as 1888 and 1894.

The affairs of each parish have, since the latter date, been managed by a Parish Council of from 5 to 15 men or women, elected by the parishioners. District Councils have charge of wider areas, and have larger powers. They are, in particular, the sanitary authorities, and are responsible for the water-supply. The County Council, whose very considerable powers extend to the whole shire, is a small parliament, which can levy rates and borrow money for public works. It manages lunatic asylums and reformatories, keeps roads and bridges in repair, controls the police in conjunction with the Quarter Sessions, appoints coroners and officers of health, and sees that the Acts relating to local government are carried out.

The Somerset County Council consists of 92 members, of whom 69 are elected every 3 years by the ratepayers of the various electoral districts, while 23 are aldermen, elected or co-opted by the 69; twelve of them in one triennial period, to serve for 6 years, and eleven in the next period, to serve for the same length of time. The

Council meets three times a year at Taunton, and twice a year at Wells. Bath, because it has a population of 50,000, is called a County Borough, and its mayor and corporation have the powers of a County Council. Somerset is a county of small towns, and possesses only 7 boroughs, that is to say, towns which have a mayor and corporation. They are Bath, Taunton, Bridgwater, Yeovil, Wells, Glastonbury, and Chard.

For the administration of justice Assizes are held at Taunton and Wells, and Quarter Sessions at Bath and Bridgwater; while Petty Sessions, presided over by local justices of the peace, are held weekly in 22 towns, to try cases and to punish those who have broken the law.

Ecclesiastical affairs are in the hands of the Bishop of Bath and Wells, the archdeacons of Bath, Taunton, and Wells, together with numerous deans and other church officials, in addition to the parish clergy.

The County Council appoints a number of Education Committees, who have charge of all Government elementary and secondary schools throughout the county.

Somerset is divided into 7 Parliamentary Divisions, known as the Northern, Wells, Frome, Eastern, Southern, Bridgwater, and Western Divisions, each of which returns one member. In addition to these Bath returns two members and Taunton one, so that the county is represented altogether by ten Members of Parliament.

24. The Roll of Honour of the County.

Few names among England's early heroes are better known than those of King Arthur and King Ina, of Alfred the Great, of Edgar the Peaceable, of Edmund Ironside, all of whom were closely associated with the history of Somerset. There may be no ground for the tradition that Christianity was introduced into Britain by Joseph of Arimathea, and that he and his companions built the first Christian church at Glastonbury; and there is certainly none for the legend that St Patrick was one of the abbots of the monastery that rose on the site of the little cell of osiers. That mistake arose through confusion with Padrig, a Welshman who took refuge there. But Dunstan, who was born at Glastonbury, and was in turn head of the Abbey and Archbishop of Canterbury, was a very real figure in Somerset history, and was one of the foremost men of his age. Many famous bishops have held the see of Bath and Wells:—Jocelyn, greatest of the builders of the cathedral; Still, the reputed author of the first English comedy; Wolsey the imperious; Laud, obstinate and ill-fated; Cranmer, Henry VIII's too willing servitor; Ken the non-juror, and author of the Morning and Evening hymns.

Of lesser dignitaries, Barclay, author of the *Shyp of Folys*, was rector of Wookey; Langhorne, the translator of Plutarch, was incumbent of Blagdon; Sydney Smith, the witty and accomplished editor of the *Edinburgh Review* held for 16 years the living of Combe Florey.

Robert Blake

(*Taunton Castle Museum*)

Out of the long list of men of letters may be named Roger Bacon, born at Ilchester in 1214, the great natural philosopher and author of the *Opus Majus*, one who as the discoverer of gunpowder and forerunner of Galileo in much of his use of the magnifying glass for scientific observation, may be said to have been born before his time; John Locke, a philosopher of another type, and author of the *Essay concerning Human Understanding*; Coryate, the quaintly facetious pedestrian traveller, author of the *Crudities Hastily gobled up in Five Moneth's Travell*, who hung up his shoes in his native church at Odcombe on returning from his journey; John Bull, organist to James I, who wrote *God save the King*; Fielding the "father of the English novel," born at Sharpham Park, near Glastonbury in 1707; Young, who deciphered the hieroglyphics on the Rosetta Stone; Norris, the early interpreter of cuneiform inscriptions; Kinglake, author of that brilliant book of Eastern travel, *Eothen*, and chronicler of the Crimean War, born near Taunton; Freeman, the historian of the Norman Conquest, who lived at Somerleaze near Wells; all of whom were born in Somerset.

Few men have left a clearer, cleaner mark in English history than Blake the indomitable, one of the greatest of England's Admirals, who served his country both by land and sea, the defender of Lyme and Taunton, the scourge of the Dutch and Spaniards. Hopton, the high-minded and gallant royalist commander who pressed him so hard at Taunton, was a Somerset man, too. Pym and Holles, two of the "Five Members," were natives of this county, although both sat for constituencies in Devon. Prynne their contemporary, the famous Puritan pamphleteer, who sacrificed his ears and his fortune in his crusade against the stage-plays of his time, was also a Somerset man. So was Dampier, buccaneer, explorer, and naturalist, one of the picturesque figures of the Stuart age, and Admiral Hood, whom Nelson called the best English officer of his time.

Other famous Somerset men are Speke, the discoverer of the great African lake Victoria Nyanza, Crosse the electrician, Quekett the microscopist, Parry the Arctic navigator, Keate the little great master of Eton, and Edward Thring the enlightened Head of Uppingham.

Somerset has produced no great poet. The *Rosamund* of Samuel Daniell is read by few except students of literature, though he wrote some excellent sonnets, and the personality of Hartley Coleridge is overshadowed by that of his distinguished father. Most of us, however, could repeat a few lines of Thomas Haynes Bayly's *Mistletoe Bough*, or *Oh, no, we never mention her*, and we may remember that, although a native of Bristol, Chatterton was born on the Somerset side of the Avon. Wordsworth and Coleridge both lived in Somerset; and the *Lyrical Ballads* of the former, and the *Ancient Mariner* of the latter were composed among the Quantock Hills. Barley Wood, the home of Hannah More, where Macaulay as a child spent many happy days, still stands near Wrington; and the memory of her untiring labours on behalf of the poor and ignorant is still green among the Mendip villages.

William Dampier

25. THE CHIEF TOWNS AND VILLAGES OF SOMERSET

(The figures in brackets after each name give the population of the parishes in 1901, and those at the end of each section are references to the pages in the text.)

SOMERSET has no large towns. Its population is scattered amongst a great number of small townships and villages and isolated farmsteads, as might be expected in a district which is almost entirely agricultural. But there are many small places or even little hamlets which are of great interest on account of their historic associations, of the beauty of their surroundings, or because they possess fine churches or manor-houses, or picturesque ruins. Geographically, part of Bristol, and especially the populous district of Bedminster, is in Somerset. But most of the city, which has been a county of itself since 1373, is in Gloucestershire.

Aller (384) is a little village four miles east of Athelney, the font in whose church may be the very one in which the Danish leader Guthrum was baptized. Here Goring's army made a stand after its defeat by Fairfax at Langport, two and a half miles away. (p. 102.)

Athelney (not a village or parish), in the heart of the flat country in the centre of Somerset, a mile from the junction of the Tone and the Parrett, was the scene of Alfred's concealment before he defeated the Danes at Ethandune, in 878. A farm stands on the spot. Nothing remains of the monastery which the king founded here. The "Alfred Jewel" was found in 1693 four miles from this spot, half-way between Athelney and Bridgwater. (pp. 21, 24, 100, 127, 145.)

Axbridge (933) is a small town on the south slope of the Mendips, where strawberries and early vegetables are grown for the market. It was an important place in Saxon times, and some very interesting old charters and other documents are preserved here. There is a fine church. (pp. 28, 44, 81.)

Banwell (1413) is a large village on the north side of the Mendips, famous for its fine and interesting church, for the old camps on the hill above it, and for the bone-caves not far off. Nothing is left of the monastery here which King Alfred gave to his friend Asser. In Banwell Abbey are traces of the ancient palace of the Bishops of Bath and Wells. (pp. 132, 134, 162.)

Bath (49,839) is the largest town in Somerset, the more important of its two cathedral cities, its only county borough, and the only town in it which is now represented by two Members of Parliament. It is most picturesquely situated on the Avon, in a sort of amphitheatre, on whose slopes its buildings rise tier above tier; and it is undoubtedly one of the most beautiful towns in the kingdom.

It has been famous, ever since the days of the Romans, for its hot medicinal springs, the hottest and most important in this country; and there is a tradition that, in times still earlier, Prince Bladud, father of the King Lear of Shakespeare's play, was cured of leprosy by bathing in its health-giving waters.

The Romans, to whom the place was known as *Aquae Solis*, or *Aquae Sulis*, perhaps after some presiding deity of the neighbourhood, took possession of it as early as 44 A.D.; it became one of their most important stations, and there have at various times been discovered in it more remarkable relics of their occupation, especially in the form of architectural remains, than have been found in any other town in England. The most striking of these are the baths, which are the largest and best preserved in the country, and the fragmentary remains of the temple of Sul Minerva, which stood on the site of the present Pump Room,

Bath and the Avon from the S.E.

Taken and burnt by Ceawlin, in 577, the town was restored by later Saxon kings, and Edgar the Peaceable was crowned there in 973, the name of the place having by that time become modified to *Bathan-ceaster*.

Bath Abbey, originally a Norman structure, was partly rebuilt in the sixteenth century, but the still-unfinished edifice was dismantled by Henry VIII's Commissioners, and long remained a roofless shell. It was roughly restored in the reign of James I, and large sums were spent, during the nineteenth century, in completing it according to the original design. It is not remarkable for architectural beauty, although it contains some fine features. But there are more monuments in it, including some by Chantrey, Flaxman, Nollekens, and Bacon, than in any other church in England except Westminster Abbey.

With the exception of the Abbey, however, the buildings of Bath are more or less modern; a large proportion of its fine squares, terraces and crescents having been designed by two architects named Wood, during the eighteenth century. The entire town, including not only its public buildings, such as the Pump Room and the Guildhall, but its shops and private houses, is built of oolite quarried in the neighbourhood, and widely celebrated as Bath stone. Bath has lent its name to many things besides this well-known building-material; Bath buns for example, Bath Oliver biscuits, Bath pipe, Bath chaps (halves of pigs' heads), Bath chairs, and Bath coating; but not, as has been shown, Bath bricks.

The most remarkable feature in the history of the town is the extraordinary popularity which it enjoyed throughout a great part of the eighteenth century, when it was the most fashionable resort in England, and was frequented by the most distinguished men and women of the time. During part of this period Bath was under the despotic rule of the adventurer Beau Nash, whose word was, for many years, law in all social affairs, and to whom the former prosperity and the present magnificence of the town are mainly owing. The golden age of Bath continued long after Nash's death, and only came to an end at the beginning of the nineteenth century, when the close of the French War made foreign travelling once more possible.

Among distinguished people who were born in Bath are Parry, the Arctic explorer, Palmer, the father of mail-coaches, and Hone, the author of the *Every-day Book*. Here, for a time, lived the two Pitts, Horace Walpole, John Wilkes. Here Burke died. In these streets Peterborough, Clive, and Nelson were once familiar figures. It was from Bath that Wolfe set out on his last campaign. It was from his house in this city that Herschel first saw the planet that, although now called Uranus, was for a time named after the King, and then after the great astronomer himself. Fielding and Smollett, Sheridan and Goldsmith, Jane Austin, Mrs Thrale, Johnson and Boswell, Gainsborough and Lawrence, Scott and Dickens;—these are among Bath's remembered visitors. (pp. 22, 42, 44, 92, 98, 122, 162, 163, 169.)

Blagdon (1089) is a village on the north side of the Mendips, once noted for its mines, and now for a great reservoir, two and a half miles long, made to supply Bristol water-works. Langhorne was rector here, and Toplady once curate-in-charge. (pp. 28, 79.)

Brean Down is a promontory at the mouth of the Axe, very interesting to the naturalist and the antiquarian. (pp. 15, 51, 53, 56—7.)

Brent Knoll is an isolated and conspicuous little hill, 457 feet high, with an ancient camp on the top, and with the fine churches of Brent and East Brent, the latter associated with Archdeacon Denison, at its foot. (pp. 40, 42, 43.)

Bridgwater (15,209), whose name is believed to be a corruption of Brugie Walter, that is, the bridge of Walter de Douai, is a borough and a river-port on the Parrett, with manufactures of Bath bricks, building-bricks, and roofing-tiles. In the parish church is a fine painting of the Descent from the Cross, said to have been taken from a Spanish privateer. In the Civil War the town was taken by Fairfax and Cromwell. The Duke of Monmouth was proclaimed king in the market-place, and it was from here that he set out on that night march which ended in his defeat at Sedgemoor, in 1685. Admiral Blake was born here, and there is a statue to his memory. (pp. 24, 86, 88 97, 107, 110, 124, 134, 146—8, 161, 164—5.)

Brockley (136), a small village nine miles south-west of Bristol, is noted for its beautiful wooded gorge called Brockley Combe. (p. 36.)

Bruton (1788), a small town in east Somerset, on the Brue, is chiefly known on account of its schools. Sexey's School was built in 1889, out of the surplus revenue of a Hospital or Almshouse established in 1638 by Hugh Sexey, who, from being a stable-boy, rose to be an Auditor to Queen Elizabeth. The Grammar School was founded by Edward VI, on the site of an older institution. The rectory stands on the site of a Benedictine Priory, later converted into an abbey of Augustinian Canons. In the church are monuments and tombs of the Berkeleys, whom it was the family custom to bury at midnight. (p. 26.)

Brympton D'Evercy (89), three miles from Yeovil, is noted for its fine Decorated church and its noble Tudor manor-house. (pp. 135, 159.)

Burnham (2897) is a breezy little watering-place on Bridgwater Bay, with a sandy beach, golf-links, a lighthouse, and a lifeboat-station. (pp. 19, 26, 49, 58, 64—5, 94.)

Burrington (417) is a picturesque village at the foot of Black Down, the highest point of the Mendips. Burrington Combe is a fine ravine, with interesting caves. Toplady wrote the hymn "Rock of Ages" while sheltering from a thunderstorm, in a rocky cleft in this gorge. (pp. 17, 35, 36.)

Cadbury. There are three ancient encampments with this name in Somerset; one near Clevedon, one near Yatton, and one, a fortress of extraordinary strength, believed to represent King Arthur's castle of Camelot, six miles from Wincanton. (pp. 55, 72, 119.)

Castle Cary (1902) is a small market-town four miles east of Evercreech, near which once stood a Norman castle. (pp. 105, 146—7.)

Castle Neroche, seven miles south of Taunton, is a very remarkable ancient earthwork, of great size and strength. Here also formerly stood a Norman castle. (pp. 119, 146—7.)

Chard (4437), in the extreme south of the county, was a place of some importance both in Roman and Saxon times, and with later historic connection with the Civil War and with the Duke of Monmouth. The lace-works employ 1000 people, and there are several fine old buildings in the town, including the Tudor court-house of the manor. (pp. 27, 44, 87.)

Charterhouse-on-Mendip (64) is a little hamlet on the top of the Mendip Hills, which was the centre of lead-mining operations from the earliest times until about forty years ago. Many Roman antiquities have been found here. The name refers to a long-vanished cell of Carthusian monks. (pp. 17, 88, 90, 99, 126, 145, 162.)

Cheddar (1975) is a large village on the south side of the Mendip Hills, which has given its name to the best kind of cheese that is made in England, and is famous for its magnificent Gorge and its beautiful stalactite caves. The tower of Cheddar church is a specially fine one, and there is a covered market-cross in the village. (pp. 17, 36, 38, 41, 81, 84, 100, 132, 134, 137.)

Cleeve Abbey, half a mile from Washford Station, is the extensive and very picturesque ruin of a monastic house of the Cistercians. (p. 144.)

Clevedon (5900) is a popular and very pretty watering-place, with fine air; but its attractions are rather in its inland scenery than its sea-shore. In the parish church is the grave of Arthur Hallam, immortalised by Tennyson in *In Memoriam*. Clevedon Court is an exceptionally fine example of domestic architecture. Coleridge brought his bride here, but of his cottage very little remains. Near the town is the ruin of a Jacobean house called Walton Castle. (pp. 54, 55, 63, 64, 82, 155.)

Compton Martin (393), on the north side of Mendip, three miles from Blagdon, is noted for its church, which contains some of the best Norman work in the county. (pp. 79, 132.)

Crewkerne (4226), a market-town on the Parrett, in the extreme south of Somerset, has some manufactures of sail-canvas, twine, and shirts. There is a fine cruciform church, with a central tower. (pp. 44, 88.)

Doulting (695), two miles from Shepton Mallet, is noted for its quarries from which, in particular, were built both Wells Cathedral and Glastonbury Abbey. Aldhelm, bishop of Sherborne, died here in 709. (pp. 44, 92, 135.)

Dulverton (1369), on the Barle near the border of Devon, in the midst of very beautiful scenery, is famous for its trout-fishing, and as a resort of Exmoor stag-hunters. Near it are Torr Steps and the ruins of Barlynch Priory. (p. 27.)

Dunster (1182) is a very picturesque village a few miles from Minehead, among some of the most beautiful scenery in the West of England. Its magnificent castle has been owned by only two families since its foundation in Norman times. Many interesting old buildings exist in the village, and there are fine stone effigies in its beautiful church. (pp. 67, 105, 108, 134, 135, 137, 146, 150, 151, 162.)

Evercreech (1188) is a large village three miles south-south-east of Shepton Mallet, with a railway junction and a fine church-tower.

Frome (11,057) is a small but busy manufacturing town in the east of Somerset, just beyond the end of the Mendip Hills, with some cloth-weaving, printing, and metal-works. Not far off are Vallis Vale, noted for its beautiful scenery, Orchardleigh, where is the only cromlech in the district, and Whatley, where is a very fine Roman tesselated pavement. (pp. 27, 87, 88.)

Glastonbury (4016), on a slight rising in the moor, once surrounded by the inland sea, and alluded to by Tennyson as the "island valley of Avilion," is a little town with most interesting legendary and historic associations connected with King Arthur and with many of the Saxon kings. Its chief features are the scanty but beautiful ruins of the abbey, the abbot's kitchen, the "George" (the ancient pilgrims' inn), the Tribunal or Court House, a large monastic barn, St John's Church, whose tower is one of the finest in Somerset, and the museum containing many of the antiquities found in the prehistoric lake-village on Godney Moor. Near the town is the Tor, on which stands the ruined church of St Michael, and Wyrral or Weary-All Hill, where Joseph of Arimathea is said to have landed, and where his staff took root and blossomed. Traditional descendants of this tree—a winter-flowering hawthorn—may be seen at the Abbey. The Abbot Dunstan was born here, and Fielding the novelist at Sharpham, not far off. (pp. 19, 40, 42, 43, 73, 76, 82, 100, 106, 132, 141—4, 155, 164.)

Highbridge (2233) is a small river-port on the Brue, with a junction for the Great Western and Somerset and Dorset Railways, with brick and tile-works, a bacon-factory, and a large timber-yard. (pp. 88, 98.)

Ilchester (433) is an ancient but now uninteresting little town on the Yeo or Ivel, in the south of Somerset. It was an important Roman station, and was the birth place of Roger Bacon. Here once stood a gaol in which in Stuart times, many hundreds of Nonconformists were cruelly imprisoned for conscience' sake, and where many died. Near the town is Lyte's Cary, now a farm, but once a fine manor-house. (pp. 137, 163.)

Ilminster (2287), a small market-town, also in the south of the county, manufactures lace, shirts, and collars. In the church are some especially fine memorial brasses, including two of the Wadhams, founders of Wadham College, Oxford. Three miles north-east of the town is Barrington Court, a famous early Tudor manor-house. (pp. 23, 80, 134.)

Kewstoke (1166) is a small village on the coast north of Weston-super-Mare, in whose church was found the Becket reliquary. Near the building is an ancient stone stair-way called the Monk's Steps. In the same parish are the picturesque ruins of Woodspring Priory, part of whose church is now a farm-house. (pp. 129, 132.)

Langport (813), a small river-port on the Parrett, is the scene of Goring's defeat by Fairfax and Cromwell in 1645. Bagehot the historian and Quekett the microscopist were born here. There is a Quekett Museum in the curious "Hanging Chapel" built over an archway. Two miles south-east of Langport are the very interesting remains of the great abbey of Muchelney, and the nearly perfect residence of its abbot. (pp. 23, 108.)

Martock (1966), five miles north-west of Yeovil, has glove and jute-matting factories. The church has a very fine timber roof; and near the building is a good fourteenth-century manor-house. (pp. 132, 137, 155, 167.)

Minehead (2511), in the far west of the county, may be called the gate of Exmoor. Once a sea-port of consequence, it is now only a watering-place. It has a fine beach, in which at low tide may be seen the remains of a submerged forest. It has an interesting church, and there is beautiful scenery in the neighbourhood. (pp. 41, 59, 60, 61, 64, 65, 94, 96, 102, 134.)

Nunney (840) is a small village about three miles west-south-west of Frome. Here is the picturesque ruin of Nunney Castle, taken by Fairfax in 1645. In the church are very interesting stone effigies of the de la Meres, former owners of the manor. (pp. 146, 149.)

Porlock (655) an ancient and picturesque village, once a port, with a very interesting history, and now a great resort of staghunters. (pp. 41, 60—62, 94, 102.)

Portishead (2544) is a small port near the mouth of the Avon with a dock and grain warehouses. A naval college has taken the place of the old training-ship "Formidable." A battery of modern guns stands on the site of a fort taken by Fairfax. (pp. 35, 54, 63, 64, 96, 126.)

Radstock (3300), half-way between Bath and Wells, is the centre of the Somerset coalfield. (pp. 40, 41, 91.)

Shepton Mallet (5200), on the south slope of the Mendips, five miles east of Wells, a once busy market-town, is now chiefly concerned with the brewing of beer. Its fine church has a magnificent panelled timber roof, and there is an old covered market-cross in the town. The Fosse Way passes near Shepton, and many Roman antiquities have been found here. Here also is the county gaol. (pp. 126, 132, 137, 163.)

Somerton (1797), a sleepy little town near the centre of the county, was the capital of the Seo-mere-saetan. Nothing remains of its Norman castle except some walling in the White Hart Inn. Here also there is a covered market-cross. (pp. 132, 146, 147.)

Stanton Drew (712) is a village two miles from Pensford, where are three remarkable stone circles. (pp. 115, 116.)

Stogumber (835), whose name is a corruption of Stoke Gomer, is a large village half-way between the Brendons and the

Quantocks, with a fine church. T vo miles away is a notable manor-house called Combe Sydenham.

Stogursey (1034), anciently Stoke Courci, is a very interesting village nine miles north-west of Bridgwater, with a large church in which are several Norman arches and some striking stone effigies. There was a Benedictine priory here and a Norman castle; some traces of both still remain. (pp. 106, 146, 147.)

Street (4018), a small town a mile south-west of Glastonbury, has an important factory of boots and shoes. In its very extensive lias quarries have been found the fossil remains of many extinct gigantic lizards. (pp. 42, 107, 163.)

Taunton (21,087), the county town of Somerset, and an important railway junction, stands on the river Tone, in a fertile plain called Taunton Dean. It has very interesting historical associations, especially with King Ina, who built the first castle here; with Perkin Warbeck, who here surrendered to Henry VII; with Blake, its heroic defender; and with the ill-fated Duke of Monmouth who was proclaimed king here and whose followers were afterwards treated with great cruelty by Kirke and Jeffreys. The restored remains of the castle now form the museum of the Somersetshire Archaeological and Natural History Society. The tower of St Mary Magdalene, re-built in 1858, is a noble example of Perpendicular architecture. Taunton is the headquarters of Prince Albert's Somerset Light Infantry, formerly the 13th regiment. (pp. 41, 80, 87, 105, 106, 107, 109, 132, 145, 146—7, 149, 164, 169.)

Uphill (518), a small port at the mouth of the Axe, was the terminus of the Roman road from Old Sarum, and there are traces of a Roman camp on the hill above it, on which stands the ruined church of St Nicholas. A cave, in which were found many bones of extinct animals, has been destroyed by quarrying. (pp. 28, 36, 55, 102, 103, 125, 162.)

Watchet (1880) is a small port in the west of the county, with a good tidal harbour and some trade. It was several times attacked, not always with success, by the Danes. (pp. 29, 42, 58, 64, 65, 91, 96, 102.)

Wellington (7300), seven miles south-west of Taunton, is noted for its large factories of serge and blankets. The Duke of Wellington took his title from this town, and there is a tall triangular column to his memory, on a hill three miles away. (pp. 41, 87, 164.)

Wells (4800), the cathedral city of Somerset, named, as Bath is, from its abundant springs, and standing at the foot of the Mendip Hills twenty miles from each of the towns of Bath, Bristol, and Bridgwater, is a strictly ecclesiastical town, with little interest apart from its architectural features. First of these is the noble cathedral, whose most striking feature, its west front, crowded with sculptured figures, has no rival. There are many interesting old buildings in the town, including the Bishop's Palace, the Deanery, the houses of the Vicars' Close, the Gates, and St Cuthbert's church, whose tower is one of great beauty. "There is no other place," wrote the historian Freeman, "where you can see so many of the ancient buildings still standing, and still put to their own use." Many famous men have been bishops of the diocese, among whom are Wolsey, Cranmer, Laud, Ken, and Still.

The historical associations of Wells are with King Henry VII, who, during his march against the followers of Perkin Warbeck, lodged at the Deanery; with Charles I, who was here during the Civil War; and with the Duke of Monmouth, the bullet-marks made by whose soldiers may still be seen on some of the figures on the west front of the cathedral. (pp. 38, 41, 91, 106, 130, 132, 134, 137, 140, 153, 155, 161, 162, 169.)

Weston-super-Mare (19,048) is a modern watering-place twenty miles from Bristol, and perhaps the most enterprising town in the county. It is well built and well kept; it has splendid sands, a long esplanade, piers, parks, golf-links, baths, and other attractions, while its mild and equable climate has made it very popular as a health-resort and as the site of many schools. Its one disadvantage is that the tide goes a very long way out, leaving exposed a vast expanse of mud. There is a small, but well-

Sidcot School

managed museum and a good library. On the hill above is Worlebury Camp, a very remarkable ancient British stronghold, once surrounded by walls of dry masonry, and containing many pits, believed to have been places of store, in which numerous very interesting antiquities have been found. (pp. 36, 55, 62, 64, 94, 109.)

Wincanton (1892), anciently Wynd-Caleton, from the stream on which it stands, was the scene of some fighting in the

Civil War, and also of a skirmish between the army of the Prince of Orange and the troops of James II. Between 1804 and 1812 about 3000 French prisoners of war were confined here. Near the town are the remains of Stavordale Priory, now a farm-house. (pp. 29, 119.)

Winscombe (1328) is a beautiful village in the heart of the Mendips, with a very fine church-tower. At Sidcot is a large and important co-educational boarding-school belonging to the Society of Friends or Quakers. (pp. 17, 35, 112, 132.)

Wookey Hole was a hyaena-den, in which were found vast quantities of bones of extinct animals. Many Roman remains have also been found here. Out of it flows the river Axe. (pp. 17, 27, 36.)

Wrington (1552), a village on the south slope of Broadfield Down, is noted for its church, whose tower Freeman considered the finest Perpendicular tower in existence. John Locke was born here, and Hannah More long lived in the neighbourhood. (pp. 79, 132.)

Yatton (1967) is a junction between the main line of the Great Western and three branches. The church has a remarkably fine west front, and contains interesting monuments. (pp. 44, 126, 132, 135.)

Yeovil (9861), one of the largest of the small towns of Somerset, on the Yeo, in the south-east of the county, has important manufactures of kid and other leather gloves. There is a striking and finely situated church, with other old buildings. Both the Great Western and the London and South Western Railway Companies have stations here. Four miles west of the town is Montacute, famous for its magnificent Elizabethan manor-house, its interesting church, and the beautiful fragment of its Cluniac priory. (pp. 23, 88, 119, 161.)

ENGLAND & WALES
37,327,479 acres

SOMERSET

Fig. 1. The Area of the "Ancient County" of Somerset (1,043,409 acres) compared with that of England and Wales

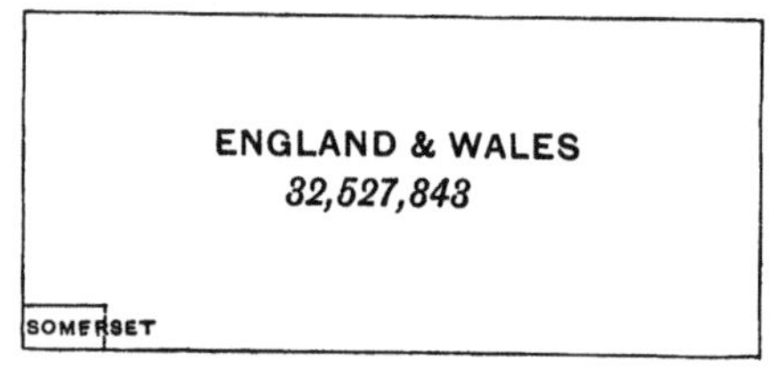

Fig. 2. The Population of Somerset (508,256) compared with that of England and Wales (in 1901)

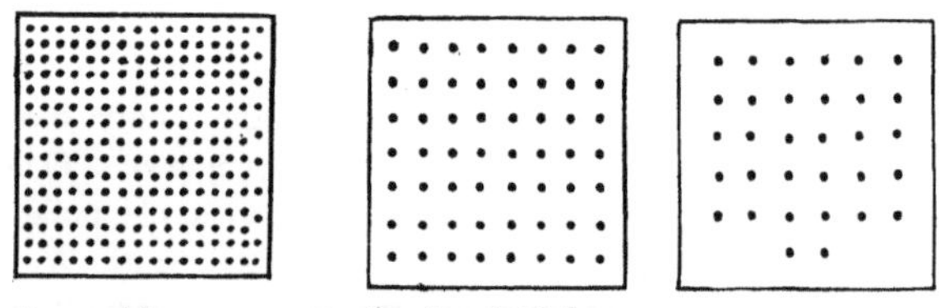

Lancashire, 2347 England and Wales, 558 Somerset, 320

Fig. 3. Comparative Density of Population to Square Mile (1901)

(*Note, each dot represents ten persons*)

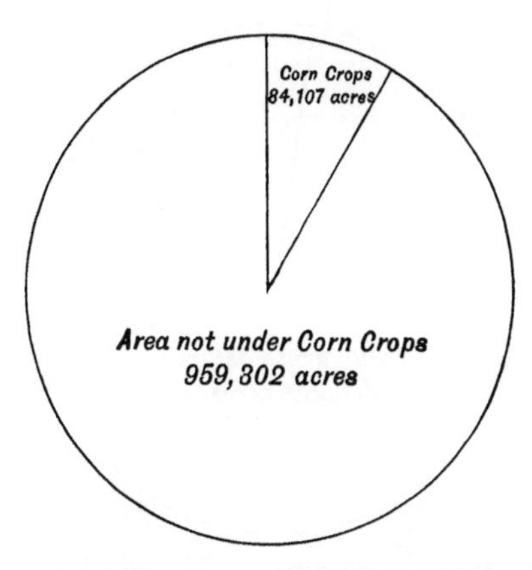

Fig. 4. Proportionate Area under Corn Crops in Somerset

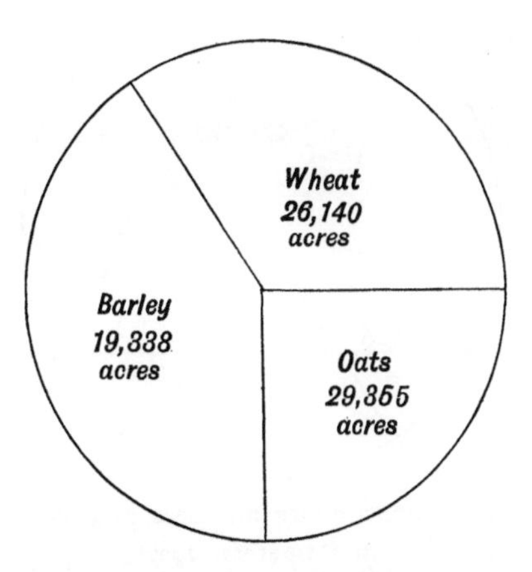

Fig. 5. Proportionate Area of chief Cereals in Somerset (1907)

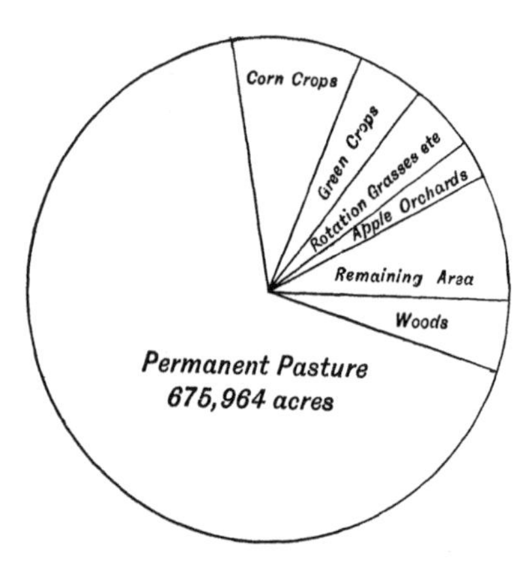

Fig. 6. Proportion of Permanent Pasture to other Areas in Somerset (1907)

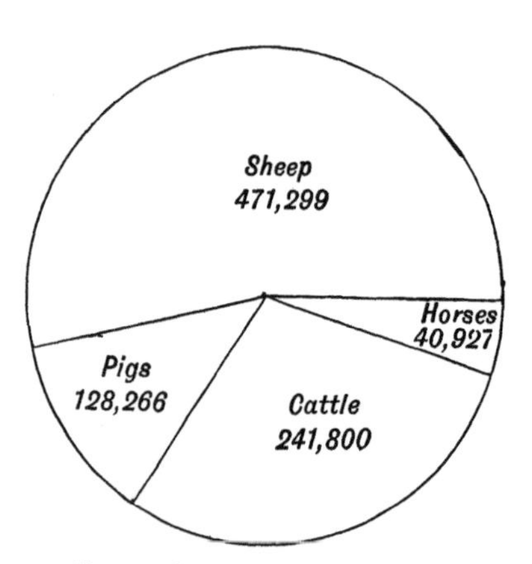

Fig. 7. Proportionate numbers of Live Stock in Somerset (1907)

www.ingramcontent.com/pod-product-compliance
Ingram Content Group UK Ltd.
Pitfield, Milton Keynes, MK11 3LW, UK
UKHW042143280225
455719UK00001B/65